Aircraft Corrosion
CONTROL GUIDE

Production Staff

Designer/Photographer Dustin Blyer
Senior Designer/Production Manager Roberta Byerly
Editor Jeff Strong

© Copyright 2007 by
Avotek Information Resources, LLC.
All Rights Reserved

International Standard Book Number 1-933189-10-X
ISBN 13: 978-1-933189-10-9
Order # T-AICO-0101

For Sale by: Avotek
A Select Aerospace Industries, Inc., company

Mail to:
P.O. Box 219
Weyers Cave, VA 24486
USA

Ship to:
200 Packaging Drive
Weyers Cave, VA 24486
USA

Toll Free: 800-828-6835
Telephone: 540-234-9090
Fax: 540-234-9399

First Edition
Second Printing
Printed in the USA

www.avotek.com

Preface

This textbook is written for the Aviation Maintenance Technology student of today. Derived from numerous sources of real world experience and research, Aircraft Corrosion Control Guide gathers that information into one book.

Corrosion control is a multifaceted area. It begins and ends with preventive measures and vigilance. In between are many steps and evaluations to determine the ultimate airworthiness of a part or airplane.

The content of this textbook does not replace the information in the manufacturer's manuals, recommendations and procedures. The procedures, techniques and practices described are only general in nature. You may not use the information in this guide for actual maintenance, repair or operations.

In order to align with today's technologies, an extensive review of the material was conducted before being included. Suggestions and ideas are always welcome. Any errors or omissions are unintentional. Please bring them to our attention. ➤

Email us at comments@avotek.com for comments or suggestions.

Avotek® Aircraft Maintenance Series

Introduction to Aircraft Maintenance
Aircraft Structural Maintenance
Aircraft System Maintenance
Aircraft Powerplant Maintenance

Avotek Avionics Series

Avionics: Fundamentals of Aircraft Electronics
Avionics: Beyond the AET
Avionics: Instruments and Auxiliary Systems
Avionics: Systems and Troubleshooting

Other Books by Avotek

Advanced Composites
Aircraft Corrosion Control Guide
Aircraft Hydraulics
Aircraft Structural Technician
Aircraft Turbine Engines
Aircraft Wiring & Electrical Installation
AMT Reference Handbook
Avotek Aeronautical Dictionary
Fundamentals of Modern Aviation
Light Sport Aircraft Inspection Procedures
Structural Composites: Advanced Composites in Aviation
Transport Category Aircraft Systems

Acknowledgments

Duncan Aviation

Evergreen Aviation

Federal Aviation Administration

National Transportation Safety Board

Select Airparts

Av-DEC®

Contents

Preface .... *iii*

Acknowledgements .... *iv*

Contents .... *v*

Chapter 1. Elements of Corrosion
    Section 1. Aircraft Corrosion *1-1*
    Section 2. Corrosion Theory *1-2*
    Section 3. Types of Corrosion *1-6*

Chapter 2. Corrosion Inspection
    Section 1. Corrosion Prevention and Control *2-1*
    Section 2. Forms of Corrosion *2-4*
    Section 3. Corrosion Inspection Procedures *2-5*
    Section 4. Corrosion Inspection Procedures for Aircraft *2-9*
    Section 5. Inspection Requirements *2-17*

Chapter 3. Corrosion Control
    Section 1. Corrosion Control Program *3-2*
    Section 2. Hazardous Materials *3-5*
    Section 3. Cleaning Procedures *3-7*
    Section 4. Basic Corrosion Removal Techniques *3-11*
    Section 5. Special Conditions for Corrosion Control *3-24*

Chapter 4. Aging Aircraft
    Section 1. Background on Aging Aircraft *4-1*
    Section 2. A Structural Integrity Program *4-5*
    Section 3. The Supplemental Structural Inspection Document *4-8*
    Section 4. Inspection Program for General Aviation Aircraft *4-11*

Chapter 5. Avionics Corrosion
    Section 1. Preventive Maintenance Program *5-1*
    Section 2. Cleaning and Preservation *5-4*
    Section 3. Corrosion Removal, Painting and Sealing *5-11*
    Section 4. Treatment of Specific Avionics Equipment *5-16*
    Section 5. Corrosion Control for Electrical Bonding/Grounding *5-27*
    Section 6. Corrosion on Electromagnetic Interference Shielding *5-31*

Index *I-1*

Chapter 1
ELEMENTS of corrosion

Section 1

Aircraft Corrosion

Corrosion and its control are of significant importance to all operators. Corrosion weakens primary structural members which, if the destruction is allowed to continue, must be replaced or reinforced in order to sustain the loads to which they may be subjected. Such replacements or reinforcements are costly and time consuming, resulting in unscheduled delays and can keep the airplane out of service for considerable periods.

Regularly scheduled preventive maintenance, when maintaining any valuable equipment, is the only sound practice. It minimizes the cost of labor expended and lost productive time. It puts both of these costs on a predictable track and removes uncertainty and guesswork as to the actual condition of the equipment. Corrosion is a natural phenomenon that destroys metal by chemical or electrochemical action, converting it into a metallic compound such as an oxide, hydroxide or sulfate. Figure 1-1-1 illustrates this principle.

The tendency of most metals to corrode creates one of the major problems in the maintenance of aircraft, particularly in areas where adverse atmospheric or weather conditions exist.

Metal corrosion is the deterioration of the metal by chemical or electrochemical attack, and the process can take place internally as well as on the surface. As in the rotting of wood, this deterioration may change the smooth surface, weaken the interior or damage or loosen adjacent parts.

If left unchecked, corrosion can cause eventual structural failure. The appearance of the corrosion varies with the metal. On aluminum alloys and magnesium, it appears as surface pitting

> **Learning Objectives:**
> - Aircraft Corrosion
> - Corrosion Theory
> - Factors Influencing Corrosion
> - Types of Corrosion
> - Pitting Corrosion
> - Filiform Corrosion
> - Concentration Cell Corrosion

Left: Today's aircraft combine many different materials in their construction.

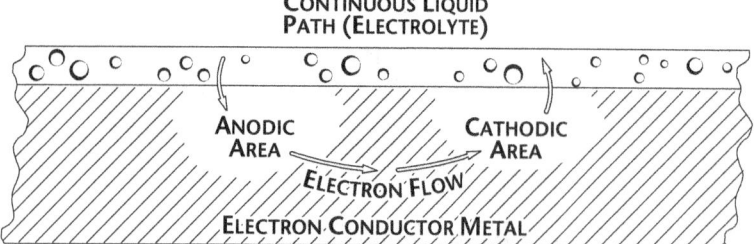

Figure 1-1-1. An electrochemical action

and etching, often combined with a gray or white powdery deposit. On copper and copper alloys, the corrosion forms a greenish film. On steel, a reddish rust is evidence of corrosion.

When the gray, white, green or reddish deposits are removed, each of the surfaces may appear etched and pitted, depending upon the length of exposure and severity of attack. If these surface pits are not too deep, they may not significantly alter the strength of the metal, however, the pits may become sites for crack development. Some types of corrosion can occur beneath surface coatings and can spread until the part fails.

Corrosion detection. The primary method of corrosion detection is visual inspection. Visual inspection must be relied upon to find corrosive attack during its incipient stage. However, many situations exist where visual inspection is not feasible, and therefore other detection techniques must be used. These other techniques include liquid dye penetrants, magnetic particle, X-ray and ultrasonic devices, all of which have achieved success in the detection of corrosion.

Visual inspections. Visual inspections of metal surfaces can reveal several signs of corrosive attack.

The most visible sign of corrosive attack is corrosion deposits. Corrosion deposits of aluminum or magnesium compounds are generally a white or grayish-white powder, while the color of ferrous compounds varies from red to dark reddish-brown.

Other indications of corrosive attack are small, localized discolorations on the surface of the metal. Surfaces protected by paint or plating may only give indications of more advanced forms of corrosive attack by the presence of blisters in the protective film, indicating that the corrosion product has a greater volume than that of the consumed metal. Bulges in lap joints may be indicative of a buildup of corrosion products, although the corrosive attack is well advanced.

Often inspection areas are obscured by structural members, equipment installations or, for some other reason, they are awkward to check visually. Magnifying glasses, mirrors and borescopes can provide the means to check obscured or difficult to reach areas. Ingenuity is encouraged, as long as the improvised inspection methods are thorough and safe.

A corrosion inspection is also difficult to accomplish if the cowling, fairings and cover plates have not been removed. Figure 1-1-2 is an example of different types of corrosion that have been missed during several successive inspections.

Figure 1-1-2. Cover plates must be removed to do a visual inspection. Failure to do so is to run the risk of missing something serious.

Section 2
Corrosion Theory

Corrosion is a natural phenomenon which attacks metal by chemical or electrochemical action and converts it into a metallic compound, such as an oxide, hydroxide, or sulfate. Corrosion is to be distinguished from erosion, which is primarily destruction by mechanical action. The corrosion occurs because of the tendency for metals to return to their natural state.

Noble metals, such as gold and platinum, do not corrode since they are chemically uncombined in their natural state. Four conditions must exist before corrosion can occur:

- Presence of a metal that will corrode (anode)

- Presence of a dissimilar conductive material (cathode) which has less tendency to corrode
- Presence of a conductive liquid (electrolyte)
- Electrical contact between the anode and cathode (usually metal-to-metal contact, or a fastener)

Elimination of any one of these conditions will stop corrosion. An example would be a paint film on the metal surface as shown in Figure 1-2-1.

Some metals, such as stainless steel and titanium, produce corrosion products that are so tightly bound to the corroding metal that they form an invisible oxide film (called a passive film) which prevents further corrosion. When the film of corrosion products is loose and porous (such as those of aluminum and magnesium), an electrolyte can easily penetrate and continue the corrosion process, producing more extensive damage than surface appearance would show.

Development of Corrosion

All corrosive attack begins on the surface of the metal. The corrosion process involves two chemical changes. The metal that is attacked or oxidized undergoes an anodic change, with the corrosive agent being reduced and undergoing a cathodic change. The tendency of most metals to corrode creates one of the major problems in the maintenance of the aircraft, particularly in areas where adverse environmental or weather conditions exist.

Paint coatings can mask the initial stages of corrosion. Since corrosion products occupy more volume than the original metal, paint surfaces should be inspected often for irregularities such as blisters, flakes, chips, and lumps.

Factors Influencing Corrosion

Some factors which influence metal corrosion and the rate of corrosion are the:

- Type of metal
- Heat treatment and grain direction
- Presence of a dissimilar, less corrodible metal (galvanic)
- Anode and cathode surface areas (in galvanic corrosion)
- Temperature
- Presence of electrolytes (hard water, salt water, battery fluids, etc.)
- Availability of oxygen

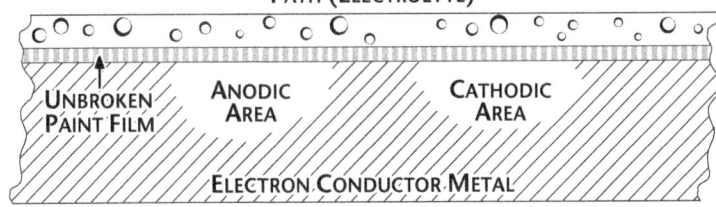

Figure 1-2-1. An unbroken paint film protects the base metal

- Presence of different concentrations of the same electrolyte
- Presence of biological organisms
- Mechanical stress on the corroding metal
- Time of exposure to a corrosive environment

Most pure metals are not suitable for aircraft construction and are used only in combination with other metals to form alloys. Most alloys are made up entirely of small crystalline regions, called grains. Corrosion can occur on surfaces of those regions which are less resistant and also at boundaries between regions, resulting in the formation of pits and intergranular corrosion. Metals have a wide range of corrosion resistance. The most active metals, such as magnesium and aluminum, corrode easily. The most noble metals, such as gold and silver, do not corrode easily.

Corrosion is accelerated by higher temperature environments which accelerate chemical reactions and allow greater moisture content at saturation in air.

Electrolytes form on surfaces when condensation, salt spray, rain, or rinse water accumulate. Dirt, salt, acidic gases, and engine exhaust gases can dissolve on wet surfaces, increasing the electrical conductivity of the electrolyte, thereby increasing the rate of corrosion.

When some of the electrolyte on a metal surface is partially confined between faying surfaces or in a deep crevice, metal in this confined area corrodes more rapidly than other metal surfaces of the same part outside this area. This type of corrosion is called an oxygen concentration cell. Corrosion occurs more rapidly than would be expected, because the reduced oxygen content of the confined electrolyte causes the adjacent metal to become anodic to other metal surfaces on the same part immersed in electrolyte exposed to the air.

Slimes, molds, fungi and other living organisms (some microscopic) can grow on damp surfaces. Once they are established, the area tends to remain damp, increasing the possibility of corrosion.

Corrosion in some cases, progresses at the same rate no matter how long the metal has been exposed to the environment. In other cases, corrosion can decrease with time, due to the barrier formed by corrosion products, or increase with time if a barrier to corrosion is being broken down.

Common Corrosive Agents

Substances that cause corrosion of metals are called corrosive agents. The most common corrosive agents are acids, alkalis, and salts. The atmosphere and water, the two most common media for these agents, may act as corrosive agents too.

Acids. In general, moderately strong acids will severely corrode most of the alloys used in airframes. The most destructive are sulfuric acid (battery acid), halogen acids (i.e., hydrochloric, hydrofluoric, and hydrobromic), nitrous oxide compounds, and organic acids found in the wastes of humans and animals.

Alkalis. Although alkalis, as a group, are generally not as corrosive as acids, aluminum and magnesium alloys are exceedingly prone to corrosive attack by many alkaline solutions unless the solutions contain a corrosion inhibitor. Particularly corrosive to aluminum are lye, potash (wood ashes), and lime (cement dust). Ammonia, an alkali, is an exception because aluminum alloys are highly resistant to it.

Salts. Most salt solutions are good electrolytes and can promote corrosive attack. Some stainless steel alloys are resistant to attack by salt solutions but aluminum alloys, magnesium alloys, and other steels are extremely vulnerable. Exposure of airframe materials to salts or their solutions is extremely undesirable.

The atmosphere. The major atmospheric corrosive agents are oxygen and airborne moisture. Corrosion often results from the direct action of atmospheric oxygen and moisture on metal, and the presence of additional moisture often accelerates corrosive attack, particularly on ferrous alloys. However, the atmosphere may also contain other corrosive gases and contaminants, particularly in industrial and marine environments, which are unusually corrosive.

Industrial atmospheres contain many contaminants, the most common of which are partially oxidized sulfur compounds. When these sulfur compounds combine with moisture, they form sulfur-based acids that are highly corrosive to most metals. In areas where there are chemical industrial plants, other corrosive atmospheric contaminants may be present in large quantities, but such conditions are usually confined to a specific locality.

Marine atmospheres contain chlorides in the form of salt particles or droplets of salt-saturated water. Since salt solutions are electrolytes, they corrosively attack aluminum and magnesium alloys which are vulnerable to this type of environment.

Water. The corrosivity of water will depend on the type and quantity of dissolved mineral and organic impurities in the water. Figure 1-2-2 shows a galley that is a prime area of corrosion. One characteristic of water which determines its corrosivity is its ability to act as an electrolyte and conduct a current. Physical factors, such as water temperature and velocity, also have a direct bearing on the corrosivity.

The most corrosive of natural waters (sea and fresh waters) are those that contain salts. Water in the open sea is extremely corrosive due to the presence of chloride ions, but waters in harbors are often even more so because they are contaminated by industrial waste.

The corrosive effects of fresh water varies from locality to locality due to the wide variety of dissolved impurities that may be present in any particular area. Some municipal waters (potable water) to which chlorine and fluorides have been added can be quite corrosive.

Figure 1-2-2. Exposure to water and food items make galley areas especially prone to corrosion.

Commercially softened water and industrially polluted rain water are usually considered to be very corrosive.

Micro-organisms

Microbial attack includes actions of bacteria, fungi, or molds. Micro-organisms occur nearly everywhere. Those organisms causing the greatest corrosion problems are bacteria and fungi.

Bacteria. Bacteria may be either aerobic or anaerobic. Aerobic bacteria require oxygen to live. They accelerate corrosion by oxidizing sulfur to produce sulfuric acid. Bacteria living adjacent to metals may promote corrosion by depleting the oxygen supply or by releasing metabolic products. Anaerobic bacteria, on the other hand, can survive only when free oxygen is not present. The metabolism of these bacteria requires them to obtain part of their sustenance by oxidizing inorganic compounds, such as iron, sulfur, hydrogen, and carbon monoxide. The resultant chemical reactions cause corrosion.

Fungi. Fungi are the growths of micro-organisms that feed on organic materials. Figure 1-2-3 is an example of corrosion caused by fungi. While low humidity does not kill microbes, it slows their growth and may prevent corrosion damage. Ideal growth conditions for most micro-organisms are temperatures between 68° and 104°F (20° and 40°C) and relative humidity between 85 and 100 percent. It was formerly thought that fungal attack could be prevented by applying moisture-proofing coatings to nutrient materials or by drying the interiors of compartments with desiccants. However, some moisture proofing coatings are attacked by mold, bacteria, or other microbes, especially if the surfaces on which they are used are contaminated.

Microbial growth occurs at the interface of water and fuel, where the fungus feeds on fuel. Organic acids, alcohols, and esters are produced by growth of the fungus. These by-products provide even better growing conditions for the fungus. The fungus typically attaches itself to the bottom of the tank and looks like a brown deposit on the tank coating when the tank is dry. The fungus growth may start again when water and fuel are present.

The spore form of some micro-organisms can remain dormant for long periods while dry, and can become active when moisture is available. When desiccants become saturated and unable to absorb moisture passing into the affected area, micro-organisms can begin to grow. Dirt, dust, and other airborne contaminants are the least recognized contributors to microbial attack. Unnoticed, small amounts of

Figure 1-2-3. Fungi contributed to the corrosion of this part

airborne debris may be sufficient to promote fungal growth.

Fungi nutrients have been considered to be only those materials that have been derived from plants or animals. Thus, wool, cotton, rope, feathers, and leather were known to provide sustenance for microbes, while metals and minerals were not considered fungi nutrients. To a large extent this rule of thumb is still valid, but the increasing complexity of synthetic materials makes it difficult or impossible to determine from the name alone whether a material will support fungus. Many otherwise resistant synthetics are rendered susceptible to fungal attack by the addition of chemicals to change the material's properties.

Damage resulting from microbial growth can occur when any of three basic mechanisms, or a combination of these, is brought into play. First, fungi are damp and have a tendency to hold moisture, which contributes to other forms of corrosion. Second, because fungi are living organisms, they need food to survive. This food is obtained from the material on which the fungi are growing. Third, these micro-organisms secrete corrosive fluids that attack many materials, including some that are not fungi nutrient.

Microbial growth must be removed completely to avoid corrosion. Microbial growth should be removed by hand with a firm nonmetallic bristle brush and water. Removal is easier if the growth is kept wet with water. Microbial growth may also be removed with steam at 100 p.s.i. and steam temperatures not exceeding 150°F (66°C). Protective clothing must be used when using steam for removing microbial growth.

From the standpoint of corrosion prevention, it is necessary to keep aircraft fuel tanks clean

Figure 1-3-1. An example of uniform surface corrosion

and use only clean, water-free fuel. Water condensate must be drained from the fuel tank frequently. Further, fuel storage facilities should be monitored to ensure that the fuel is clean.

Biocide treatment may be used for control of microorganisms in jet fuel tanks. Complete elimination of water and contaminants from the fuel all but prohibits tank corrosion; unfortunately, keeping a jet fuel tank totally free of water and debris isn't easily accomplished. Experience with fuel containing biocide has shown that, when used in the correct proportions, it is effective in eliminating many types of microbial growth, thus reducing tank corrosion.

Metallic Mercury Corrosion on Aluminum Alloys

Spilled mercury on aluminum should be cleaned immediately because mercury causes corrosion attack which is rapid in both pitting and intergranular attack and is very difficult to control. The most devastating effect of mercury spillage on aluminum alloys is the formation of an amalgam which proceeds rapidly along grain boundaries, causing liquid metal embrittlement. If the aluminum alloy part is under tension stress, this embrittlement will result in splitting with an appearance similar to severe exfoliation. X-ray inspection may be an effective method of locating the small particles of spilled mercury because the dense mercury will show up readily on the X-ray film.

Section 3

Types Of Corrosion

There are many different types of corrosive attack and these will vary with the metal concerned, corrosive media location, and time exposure. For descriptive purposes, the types are discussed under what is considered the most commonly accepted titles.

Forms of Corrosion

There are many forms of corrosion. The form of corrosion depends on the metal involved and the corrosion-producing agents present.

Oxidation. One of the most simple forms of corrosion, and perhaps the one with which we are most familiar, is dry corrosion, or oxidation.

When aluminum is exposed to a gas containing oxygen, a chemical reaction takes place at the surface between the metal and the gas. In this case, two aluminum atoms join three oxygen atoms to form aluminum oxide: AL_2O_3.

If the metal is iron or steel (ferrous metal), two atoms of iron will join three atoms of oxygen and form iron oxide or rust: Fe_2O_3.

There is one big difference between iron oxide and aluminum oxide. If the film of aluminum oxide is unbroken, further reaction with the oxygen continues at a greatly reduced rate; indeed, it almost stops. Iron oxide, on the other hand, forms a porous or interrupted film, and the metal will continue to react with the oxygen in the air until the metal is completely eaten away.

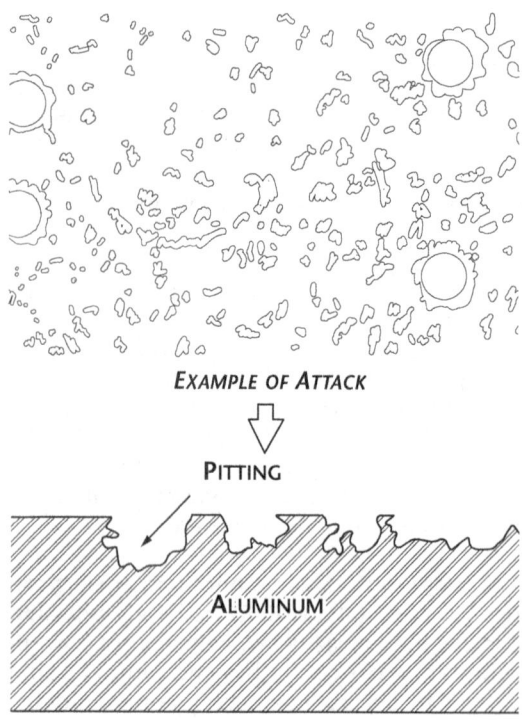

Figure 1-3-2. Pitting, as seen on the surface and in a magnified cross-section

Figure 1-3-3. This magnesium casting was installed in a passenger door. Pitting corrosion formed on the lower edge of the gasket (removed) when the sealant became worn. The pitting is more than 0.1 inch deep.

Uniform surface corrosion. Surface corrosion, as shown in Figure 1-3-1, appears as a general roughening, etching or pitting of the surface of a metal, frequently accompanied by a powdery deposit of corrosion products. Surface corrosion may be caused by either direct chemical or electrochemical attack.

Occasionally, corrosion will spread under the surface coating and cannot be recognized by either a roughening of the surface or a powdery deposit. Instead, the paint or plating will be lifted off the surface in small blisters, which result from the pressure of the underlying accumulation of corrosion products.

A common type of uniform surface corrosion is caused by the reaction of metallic surfaces with airborne chlorine or sulphur compounds, oxygen, or moisture in the atmosphere. Often combinations of these agents may attack a surface simultaneously. Reactive compounds from exhaust gases, as well as fumes from storage batteries, frequently cause uniform surface corrosion.

The amount of damage caused by uniform surface corrosion is ordinarily determined by comparing the thickness of the corroded metal with that of an undamaged specimen.

On a polished surface, this type of corrosion is first seen as a general dulling of the surface, and if the attack is allowed to continue, the surface becomes rough and possibly frosted in appearance. The discoloration or general dulling of metal created by exposure to elevated temperatures is not to be considered as uniform etch corrosion.

Pitting corrosion. Pitting corrosion is confined to very small areas of the metal surface, while the remainder of the surface is unaffected. The corrosion pits are often randomly located over the surface, however, some preferential attack may occur at the grain boundaries of the metal. A surface and cross-sectional example are shown in Figure 1-3-2.

Pitting results from the chemical action of moisture, acid, alkali or saline solutions on the metal after the paint, surface oxide, or other protective film has either been removed or penetrated. Once pitting has begun, it is propagated by means of concentration cells or galvanic action.

The pits found in pitting types of corrosion usually have a rather short, well-defined edge with walls that run almost perpendicular to the surface of the metal.

It is first noticeable as a white or gray powdery deposit, similar to dust, which blotches the surface. When the deposit is cleaned away, tiny pits or holes can be seen in the surface. Pitting corrosion may also occur in other types of metal alloys. The combination of small active anodes to large passive cathodes causes severe pitting. The principle also applies to metals which have been passivated by chemical treatments, as well as for metals which develop passivation due to environmental condition.

All forms of pits have one thing in common regardless of their shape: they penetrate deeply into the metal, and cause damage completely out of proportion to the amount of metal consumed. A good example is Figure 1-3-3.

Intergranular corrosion. Intergranular corrosion concentrates on the boundaries of the metal grains, first consuming the material between the grain boundaries and then attacking the grains themselves.

Each grain has a clearly defined boundary which, from a chemical point of view, differs from the metal within the grain center. The grain boundary and grain center can react with each other as anode and cathode when in contact with an electrolyte. Rapid selective corrosion at the grain boundary can occur with subsequent delamination. High-strength aluminum alloys such as 2014 and 7075 are more susceptible to intergranular corrosion if they have been improperly heat-treated and are then exposed to a corrosive environment.

The damage from intergranular corrosion, like pitting corrosion, causes a loss of strength and ductility that is out of proportion to the amount of the metal destroyed. An example of intergranular corrosion is shown in Figure 1-3-4.

Aluminum alloys and some stainless steels are particularly susceptible to intergranular corrosion. A lack of uniformity in the grain of these metals is caused by changes that occur in the alloy during heating and cooling.

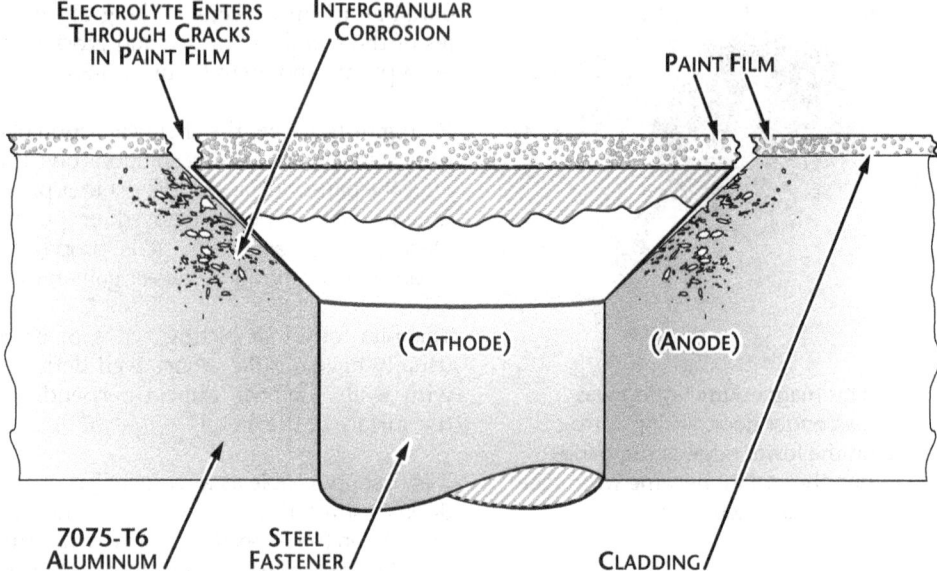

Figure 1-3-4. Intergranular corrosion of 7075-T6 aluminum, adjacent to a steel fastener

Intergranular corrosion may exist without visible surface evidence, and is difficult to detect in its early stages. Ultrasonic and eddy current inspection methods provide the best success in detecting this particular form of corrosion.

Exfoliation. Exfoliation is a severely destructive form of intergranular corrosion, characterized by the actual leafing-out of corroded sections of metal away from the rest of the part, as can be seen in Figure 1-3-5.

Exfoliation is most prone to occur in wrought products such as extrusions, thick sheet, thin plate and certain die-forged shapes which have a thin, highly elongated platelet type grain structure. This is in contrast with other wrought products and cast products that tend to have an equal axis grain structure.

This is because the extrusion process elongates the grains of the metal. The corrosive attack on the grain boundary material produces corrosion products that take up more volume than that originally occupied by the unaffected grain boundaries, causing the part to swell.

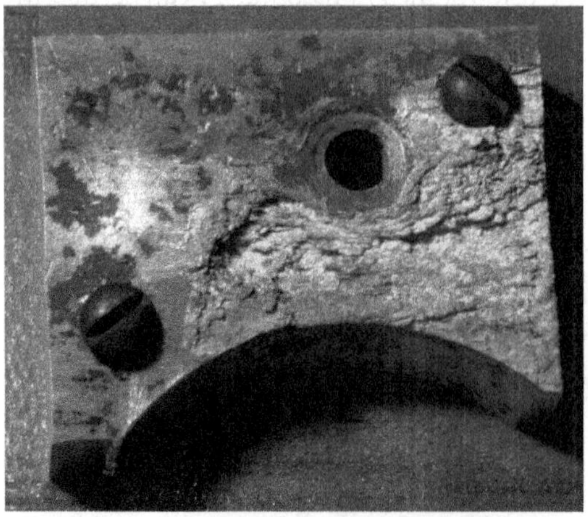

By the time exfoliation corrosion is detected, the intergranular attack usually is so advanced that the static strength of the part is impaired because of the reduction of its effective cross-sectional area.

Galvanic corrosion. Galvanic cells may originate from localized differences of materials in the surface of an alloy, because the dissimilar metals in alloys provide a basis for galvanic action within the alloys themselves and is illustrated in Figure 1-3-6.

If an electrolytic medium is provided (like the condensation from a salt-air atmosphere), the metal can literally destroy itself.

The degree of attack depends on the relative activity of the two surfaces; the greater the difference in activity, the more severe the attack.

Concentration Cell Corrosion

This type of corrosion occurs when two or more areas of metal surface are in contact with different concentrations of the same solution. This is shown in Figure 1-3-7. Concentration cell corrosion is corrosion of metals in a metal-to-metal joint, corrosion at the edge of a joint even though joined metals are identical, or corrosion of a spot on the metal surface covered by a foreign material. Another term for this type of corrosion is crevice corrosion. Metal ion concentration cells, oxygen concentration cells, and

Figure 1-3-5. In this example the exfoliation is so advanced that the part is scrap.

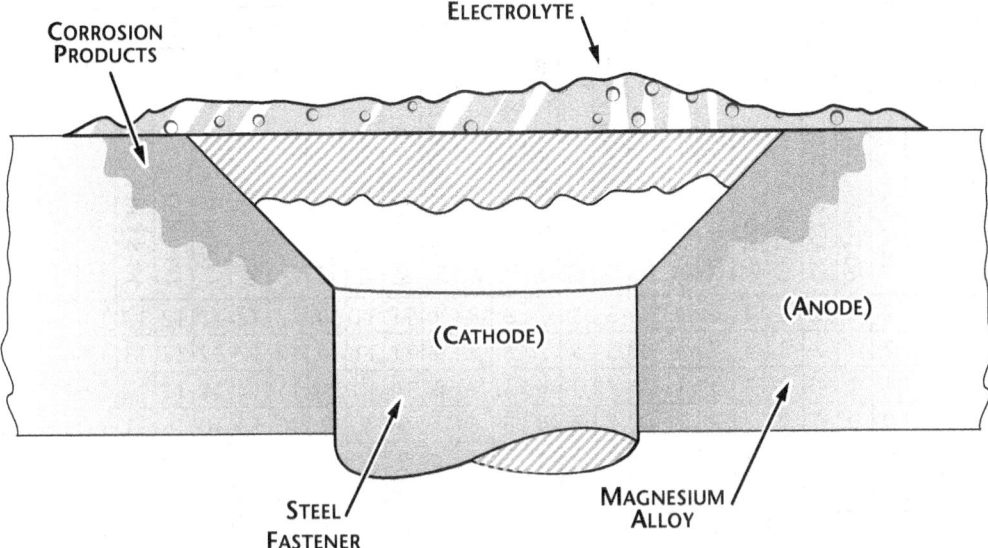

Figure 1-3-6. Galvanic cell corrosion

active-passive cells are the three general types of concentration cell corrosion.

Metal ion concentration cells. With metal ion concentration cell corrosion, the solution may consist of water and ions of the metal which is in contact with the water. A high concentration of the metal ions will normally exist under faying surfaces where the solution is stagnant, and a low concentration of metal ions will exist adjacent to the crevice which is created by the faying surface. An electrical potential will exist between the two points; the area of the metal in contact with the high metal ion concentration will be anodic and will be corroded.

If water is allowed to stagnate in a metallic structure, either an oxygen or metal ion concentration cell can develop. Small deposits of corrosion products may give an indication of the damage that has been sustained.

Oxygen concentration cells. The solution in contact with the metal surface will normally contain dissolved oxygen. An oxygen cell can develop at any point where the oxygen in the air is not allowed to diffuse into the solution, thereby creating a difference in oxygen concentration between two points. Typical locations of oxygen concentration cells are under either metallic or nonmetallic deposits on the metal surface and under faying surfaces such as riveted lap joints. Oxygen cells can also develop under gaskets, wood, rubber, and other materials in contact with the metal surface.

Corrosion will occur at the area of low oxygen concentration (anode). Alloys, such as stainless steel, which owe their corrosion resistance to surface passivity, are particularly susceptible to this type of crevice corrosion. Figure 1-3-8 is a casting that has been destroyed by corrosion.

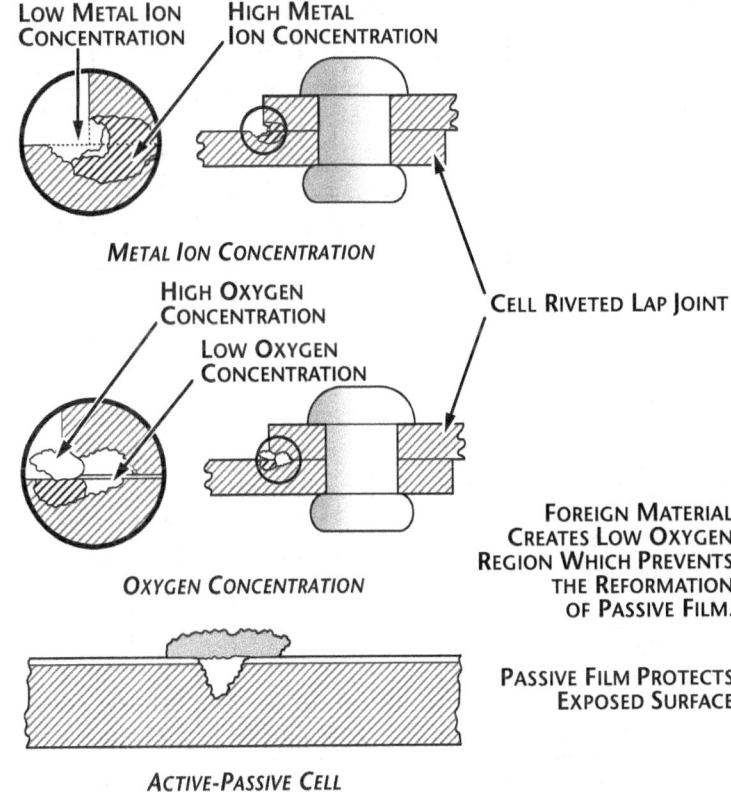

Figure 1-3-7. Concentration cell corrosion

Filiform corrosion. Metals coated with organic coatings tend to undergo a type of corrosion resulting in numerous threadlike filaments of corrosion products under the coating. It is caused by the diffusing of oxygen and water through the coating and is considered a special type of oxygen concentration cell corrosion.

One of the reasons for filiform corrosion has been the mistake of top coating a wash primer

		CORRODED END (ANODIC OR LEAST NOBLE)												PROTECTED END (CATHODIC OR MOST NOBLE)																
	GROUP NUMBER	1	1	2	3	3	3	3	4	5	5	5	5	6	6	8	9	9	10	12	12	12	13	13	13	13	14	14	14	14	14
GROUP NUMBER	METAL OR ALLOY	MAGNESIUM	MAGN. ALLOY	ZINC	1100	3003	6061-T6	CLAD ALLOYS	CADMIUM	2017-T4	2014-T4	2024-T4	7075-T6	STEEL WROUGHT	STEEL CAST	50-50 SOLDER	LEAD	TIN	MANG. BRONZE	BRASSES	ALUM. BRONZE	COPPER	NICKEL	INCONEL	TYPE 410	TYPE 431	18-8 CRES	TITANIUM	MONEL	SILVER	GRAPHITE
1	MAGNESIUM	0	0	1	2	2	2	2	3	4	4	4	4	5	5	7	8	8	9	11	11	11	12	12	12	12	13	13	13	13	13
1	MAGN. ALLOY		0	1	2	2	2	2	3	4	4	4	4	5	5	7	8	8	9	11	11	11	12	12	12	12	13	13	13	13	13
2	ZINC			0	1	1	1	1	2	3	3	3	3	4	4	6	7	7	8	10	10	10	11	11	11	11	12	12	12	12	12
3	1100				0	0	0	0	1	2	2	2	2	3	3	5	6	6	7	9	9	9	10	10	10	10	11	11	11	11	11
3	3003					0	0	0	1	2	2	2	2	3	2	5	6	6	7	9	9	9	10	10	10	10	11	11	11	11	11
3	6061-T6						0	0	1	2	2	2	2	3	2	5	6	6	7	9	9	9	10	10	10	10	11	11	11	11	11
3	CLAD ALLOYS							0	1	2	2	2	2	3	2	5	6	6	7	9	9	9	10	10	10	10	11	11	11	11	11
4	CADMIUM								0	1	1	1	1	2	2	4	5	5	6	8	8	8	9	9	9	9	10	10	10	10	10
5	2017-T4									0	0	0	0	1	1	3	4	4	5	7	7	7	8	8	8	8	9	9	9	9	9
5	2014-T4										0	0	0	1	1	3	4	4	5	7	7	7	8	8	8	8	9	9	9	9	9
5	2024-T4											0	0	1	1	3	4	4	5	7	7	7	8	8	8	8	9	9	9	9	9
5	7075-T6												0	1	1	3	4	4	5	7	7	7	8	8	8	8	9	9	9	9	9
6	STEEL WROUGHT													0	0	2	3	3	4	6	6	6	7	7	7	7	8	8	8	8	8
6	STEEL CAST														0	2	3	3	4	6	6	6	7	7	7	7	8	8	8	8	8
8	50-50 SOLDER															0	1	1	2	4	4	4	5	5	5	5	6	6	6	6	6
9	LEAD																0	0	1	3	3	3	4	4	4	4	5	5	5	5	5
10	TIN																	0	1	3	3	3	4	4	4	4	5	5	5	5	5
12	MANG. BRONZE																		0	2	2	2	3	3	3	3	4	4	4	4	4
12	BRASSES																			0	0	0	1	1	1	1	2	2	2	2	2
12	ALUM. BRONZE																				0	0	1	1	1	1	2	2	2	2	2
13	COPPER																					0	1	1	1	1	2	2	2	2	2
13	NICKEL																						0	0	0	0	1	1	1	1	1
13	INCONEL																							0	0	0	1	1	1	1	1
13	TYPE-410																								0	0	1	1	1	1	1
14	TYPE 431																									0	1	1	1	1	1
14	18-8 CRES																										0	0	0	0	0
14	TITANIUM																											0	0	0	0
14	MONEL																												0	0	0
14	SILVER																													0	0
14	GRAPHITE																														0

Note: The larger the number, the greater the tendency for galvanic corrosion.

Table 1-3-1. Galvanic grouping of metals — the smaller the number, the less active

with an epoxy or polyurethane system before the acids in the primer where completely converted. Any ambient moisture that gets under the finish is then trapped and begins to react with the acids and causes filiform corrosion. Figure 1-3-9 shows the resulting corrosion.

Filiform occurs when the relative humidity of the air is between 78 and 90 percent and the surface is slightly acidic. This corrosion usually attacks steel and aluminum surfaces. The characteristic worm like traces never cross on steel, but they will cross under one another on aluminum which makes the damage deeper and more severe for aluminum. If the corrosion is not removed, the area treated, and a protective finish applied, the corrosion can lead to inter-granular corrosion, especially around fasteners and at seams. Filiform corrosion can be removed using glass bead blasting material with portable abrasive blasting equipment or sanding. Filiform corrosion can be prevented by storing aircraft in an environment with a relative humidity below 70 percent, using coating systems having a low rate of diffusion for oxygen and water vapors, and by washing the aircraft to remove acidic contaminants from the surface.

Active-passive cells. Metals which depend on a tightly adhering passive film, usually an oxide for corrosion protection, such as corrosion resistant steel, are prone to rapid corrosive attack by active-passive cells. The corrosive action usually starts as an oxygen concentration cell. As an example, salt deposits on the metal surface in the presence of water containing oxygen can create the oxygen cell. The passive film will be broken beneath the dirt particle. Once the passive film is broken, the active metal beneath the film will be exposed to corrosive attack. An electrical potential will develop between the large area of the cathode (passive film) and the small area of the anode (active metal). Rapid pitting of the active metal will result.

This reaction can become locally intense. The reaction is augmented by the affected area, since the proportion of the exposed base metal is small compared to the surrounding non-reactive metal. This effectively concentrates the focal point of the reaction, often resulting in deep pits in a short time and a greater rate of corrosion.

Of the materials listed in Table 1-3-1, some items are quite active and corrode easily. They require maximum protection. Other materials that are close together numerically are the least active, therefore requiring minimum protection.

The rate which corrosion occurs depends on the difference in the activities. The greater the difference in activity, the faster corrosion occurs. For example, magnesium would corrode very quickly when coupled with gold in a humid atmosphere, but aluminum would corrode very slowly in contact with cadmium. The rate of galvanic corrosion also depends on the size of the parts in contact. If the surface area of the corroding metal (the anode) is smaller than the surface area of the less active metal (the cathode), corrosion will be rapid and severe. When the corroding metal is larger than the less active metal, corrosion will be slow and superficial. For example, an aluminum fastener in contact with a relatively inert Monel structure may corrode severely, while a Monel bracket secured to a large aluminum member would result in a relatively superficial attack on the aluminum sheet.

Corrosion and Mechanical Factors

Corrosive attack is often aggravated by mechanical factors that are either within the part (residual) or applied to the part (cyclic service loads). Erosion by sand and/or rain and mechanical wear will remove surface protective films and contribute to corrosive attack of underlying metal surfaces. Corrosive attack that is aided by some mechanical factor usually causes the part to degenerate at an accelerated rate compared to the rate at which the same part would deteriorate if it were subjected solely to corrosive attack. Environmental conditions and the composition of the alloy also influence the extent of attack. Examples of this kind of alliance are stress-corrosion cracking, corrosion fatigue, and fretting corrosion.

Stress corrosion. Stress corrosion occurs as the result of the combined effect of sustained tensile stresses and a corrosive environment. While stress corrosion cracking is found in most metal systems, it is particularly characteristic of aluminum, copper, certain stainless steels and high-strength alloy steels (more than 240,000 p.s.i.). It usually occurs along lines of cold-working and may be trans-granular or intergranular in nature. Aluminum alloy bellcranks with pressed-in bushings, landing gear shock struts with pipe-thread type grease fittings, clevis pin joints, shrink fits and overstressed tubing B-nuts are examples of parts which are susceptible to stress corrosion cracking which is evident in Figure 1-3-10.

Figure 1-3-8. This casting developed extensive corrosion when the built-up sealant cracked and allowed water to contact the metal. The part appeared sound until corrosion extended beyond the sealant.

Figure 1-3-9. Filiform corrosion is a form of oxygen concentration cell corrosion that occurs under painted surfaces.

Figure 1-3-10. Stress corrosion cracking

Crack initiation generally results from a physical breakdown of protective surface films and the subsequent corrosive attack on the part, in conjunction with the application of stress forces.

Several elements determine crack propagation. Factors such as the alloy type, changes in the composition of the alloy or its environment, the type of heat treatment and the method of metal forming all contribute to the direction and length of the crack. Finally, for each type alloy, specific environmental conditions must exist before stress corrosion can occur.

Internal stress. Internal stress may be trapped in a part of structure during manufacturing processes such as cold working or by unequal cooling from high temperatures. Most manufacturers follow up these processes with a stress relief operation. Even so, sometimes stress remains trapped. The stress may be externally introduced in part structure by riveting, welding, bolting, clamping, press fit, etc. If a slight mismatch occurs, or a fastener is over-torque, internal stress will be present.

Internal stress is more important than design stress, because stress corrosion is difficult to recognize before it has overcome the design safety factor. The level of stress varies from point to point within the metal. Stresses near the yield strength are generally necessary to promote stress corrosion cracking. However, failures may occur at lower stresses. Specific environments have been identified which cause stress corrosion cracking of certain alloys.

- Salt solutions and sea water cause stress corrosion cracking of high-strength, heat-treated steel and aluminum alloys.

- Methyl alcohol-hydrochloric acid solutions will cause stress corrosion cracking of some titanium alloys.

- Magnesium alloys may stress corrode in moist air.

- Stress corrosion may be reduced by applying protective coatings, stress relief heat treatments, using corrosion inhibitors, or controlling the environment.

Stress-corrosion cracking. Stress-corrosion cracking is an intergranular cracking of the metal which is caused by a combination of stress and corrosion. Stress may be caused by internal or external loading. Internal stresses are produced by nonuniform deformation during cold working, by unequal cooling from high temperatures, and by internal structural rearrangement involving volume changes shown in Figure 1-3-11. Internal stresses are induced when a piece of structure is deformed during an assembly operation, (i.e., during pressing in bushings, shrinking a part for press fit, installing interference bolts, installing rivets, etc.).

Shot peening a metal surface increases resistance to stress corrosion cracking by creating compressive stresses on the surface which should be overcome by applied tensile stress before the surface sees any tension load. Therefore, the threshold stress level is increased.

Corrosion fatigue. Corrosion fatigue is caused by the combined effects of cyclic stress and corrosion. No metal is immune to some reduction in its resistance to cyclic stressing if the metal is in a corrosive environment. Damage from corrosion fatigue is greater than the sum of the damage from both cyclic stresses and corrosion. Corrosion fatigue failure occurs in two stages. During the first stage, the combined action of corrosion and cyclic stress damages the metal by pitting and crack formation to such a degree that fracture by cyclic stressing will ultimately occur, even if the corrosive environment is completely removed.

The second stage is essentially a fatigue stage in which failure proceeds by propagation of the crack (often from a corrosion pit or pits) and is controlled primarily by stress concentration effects and the physical properties of the metal. Fracture of a metal part, due to fatigue corrosion, generally occurs at a stress level far below the fatigue limit in laboratory air, even though the amount of corrosion is relatively small. For this reason, protection of all parts subject to alternating stress is particularly important, even in environments that are only mildly corrosive.

Fretting corrosion. Fretting corrosion is a particularly damaging form of corrosive attack that occurs when two mating surfaces, normally at rest with respect to one another, are subject to slight relative motion. While the fit between two surfaces may be very tight, it is rarely tight enough to prevent oxygen or other corrosive agents from entering and attacking unprotected surfaces. Figure 1-3-12 is an example of fretting corrosion, commonly called smoking rivets, often found on engine cowls and wing skins.

Mechanical fretting and chemical corrosion in combined action is referred to as fretting corrosion. It is characterized by pitting of the surfaces and the generation of considerable quantities of finely divided debris. Since the restricted movements of the two surfaces prevent the debris from escaping very easily, an extremely localized abrasion occurs.

The protective film on the metallic surfaces is removed by this rubbing action. With continued rubbing, metal particles sheared from the surface of the metal combine with oxygen to form metal oxide. As these oxides accumulate, they cause damage by abrasive action and increased local stress. This is one corrosion reaction that is not driven by an electrolyte, and in fact, moisture may inhibit the reaction. Application of a lubricant or installation of a fretting-resistant material between the two surfaces can reduce fretting corrosion.

The presence of water vapor greatly increases this type of deterioration. If the contact areas are small and sharp, deep grooves resembling Brinell markings or pressure indentations may be worn in the rubbing surface. As a result, this type of corrosion (on bearing surfaces) has also been called false Brinelling.

Heat treatment. Heat treatment of airframe materials should be rigidly controlled to maintain their corrosion resistance as well as to improve their essential mechanical properties. For example, improper heat treatment of clad aluminum alloy may cause the cladding to incur excessive diffusion because the solution heat treatment is too long or at too high a temperature. This degrades the inherent resistance of the cladding itself, and reduces its ability to provide protection to the core aluminum alloy.

Aluminum alloys which contain appreciable amounts of copper and zinc are highly vulnerable to intergranular corrosion attack if not quenched rapidly during heat treatment or other special treatment. Stainless steel alloys are susceptible to carbide sensitization when slowly cooled after welding or high temperature heat treatment. Post-weld heat treatments are normally advisable for reduction of residual stress.

Hydrogen embrittlement. Environmentally induced failure processes may often be the result of hydrogen damage rather than oxidation. Atomic hydrogen is a cathodic product of many electrochemical reactions, forming during naturally occurring corrosion reactions as well as during many plating or pickling processes. Whether hydrogen is liberated as a gas, or atomic hydrogen is absorbed by the metal, depends on the surface chemistry of the metal.

Atomic hydrogen due to its small size and mass, has very high diffusivity in most metals. It will therefore penetrate most clean metal surfaces easily and migrate rapidly to favorable sites where it may remain in solution, precipitate as molecular hydrogen to form small pressurized cavities, cracks or large blisters, or it may react with the base metal or with alloying elements to form hydrides.

The accumulation of hydrogen in high strength alloys often leads to cracking, and this often occurs in statically loaded components several hours or even days after the initial application of the load or exposure to the source of hydrogen. Cracking of this type is often referred to as hydrogen stress cracking, hydrogen delayed cracking, or hydrogen induced cracking. Similar fracture processes can occur in new and unused parts when heat treatments or machining have left residual stresses in the parts, and have then been exposed to a source of hydrogen. For this reason, all processes such as pickling or electroplating must be carried out under well controlled conditions to minimize the amount of hydrogen generated.

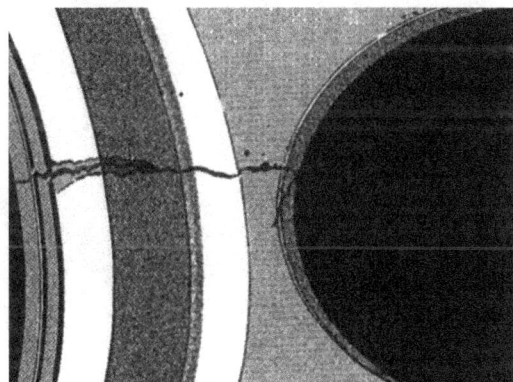

Figure 1-3-11. A typical example of cracking caused by stress corrosion

Figure 1-3-12. Fretting corrosion

Chapter 2

CORROSION inspection

Section 1

Corrosion Prevention and Control

Corrosion is the single greatest threat to the integrity of airframe structures. It is a natural chemical or electrochemical process that occurs on metals and metal alloys. Corrosion weakens the metals it attacks and, if left untreated, will cause metal parts to fail.

A number of factors promote the spread of corrosion. These include air pollution, the type of metal used in components, or the type of stress the metal is subjected to. However, unless a metal part suffers direct chemical attack by acid or a mercury spill, corrosion will only occur in the presence of water and oxygen.

Factors in Corrosion Control

Corrosion factors. The degree of severity, cause, and type of corrosion depend on many factors, including the size or thickness of the part, the material, heat treatment of the material, protective finishes, environmental conditions, preventive measures, and design.

Thick structural sections are more susceptible to corrosive attack because of variations in their composition, particularly if the sections have been heat-treated during fabrication. When a large section is machined or chem-milled after heat treatment, the corrosion characteristics of thinner sections may be different from those of thicker areas. Section size is based on structural requirements and cannot be changed for the purpose of controlling corrosion. From a maintenance standpoint, the correct approach is

Learning Objectives:

- Corrosion Prevention
- Corrosion Inspection
- Inspection Procedures
- Inspection Requirements

Left. **Proper inspection for corrosion requires extensive disassembly of the airplane**

CONTACTING METALS	ALUMINUM ALLOY	CADMIUM PLATE	ZINC PLATE	CARBON AND ALLOY STEELS	LEAD	TIN COATING	COPPER AND ALLOYS	NICKEL AND ALLOYS	TITANIUM ALLOYS	CHROMIUM PLATE	CORROSION RESISTING STEEL	MAGNESIUM ALLOYS
Aluminum Alloy				▓	▓	▓	▓	▓	▓	▓	▓	▓
Cadmium Plate				▓	▓	▓	▓	▓	▓	▓	▓	▓
Zinc Plate				▓	▓	▓	▓	▓	▓	▓	▓	▓
Carbon and Alloy Steels	▓	▓	▓				▓	▓	▓	▓	▓	▓
Lead	▓	▓	▓				▓	▓	▓	▓	▓	▓
Tin Coating	▓	▓	▓				▓	▓	▓	▓	▓	▓
Copper and Alloys	▓	▓	▓	▓	▓	▓						▓
Titanium and Alloys	▓	▓	▓	▓	▓	▓						▓
Chromium Plate	▓	▓	▓	▓	▓	▓						▓
Corrosion Resisting Steel	▓	▓	▓	▓	▓	▓						▓
Magnesium Alloys	▓	▓	▓	▓	▓	▓	▓	▓	▓	▓	▓	
DARKER SHADED AREAS INDICATE DISSIMILIAR METAL CONTACT												

Table 2-1-1. Differences in electrode potential among various structural materials.

one of recognizing the need to ensure the integrity and strength of major structural parts and maintaining permanent protection over such areas at all times.

In-service stresses and field repairs may affect the rates and types of corrosion. Aircraft structure under high cyclic stresses, such as helicopter main rotors, are particularly subject to stress-corrosion cracking. Areas adjacent to weld repaired items often have corrosion due to insufficient removal of the weld flux, or, for some steels, buildup of a magnetic field. Areas such as these should be closely inspected for signs of corrosion and, treated accordingly.

Corrosion Control in Design

Corrosion control as a means to minimize corrosion of an aircraft should be introduced during the design phase, since corrosion is the deterioration of metals resulting from reactions between metals and their environment.

The nature of the material is a fundamental factor in corrosion. High-strength, heat-treatable aluminum and magnesium alloys are very susceptible to corrosion, while titanium and some stainless steel alloys are less susceptible in atmospheric environment. The aircraft manufacturer selects material for the aircraft based on material strength, weight, and cost, while corrosion resistance is often a secondary consideration. However, corrosion control should be considered as early as possible during the preliminary design phase.

The use of corrosion-resistant materials in any design normally involves additional weight to achieve the required strength. Since weight consideration is a major factor in the construction of airframes, the primary means of preventing corrosion is by the use of protective coatings and proper maintenance procedures.

The use of corrosion-resistant alloys is not a cure-all for corrosion prevention. A common mistake is to replace a corroded part with a corrosion-resistant alloy only to find that the corrosion has now shifted to another part and increased in severity.

Corrosion problems are minimized if the material to be protected is intrinsically resistant to corrosion. Aluminum copper alloys are known to have better stress-corrosion resistance and better fatigue strength properties than aluminum zinc alloys; therefore, they are often used as primary structural materials.

Galvanic corrosion is created by dissimilar metals being in contact with each other. The galvanic series of metals and alloys, shown in Table 2-1-1, is a factor that should be considered in the repair of aircraft. The further apart the metals listed in

Table 2-1-2 are, the greater the tendency will be for galvanic corrosion. The metals grouped together in Table 2-1-2 have little differences in electrical potential; thus they are relatively safe to use in contact with one another. However, the coupling of metals from different groups will result in corrosion of the group having a lower number.

Geographical Location and Environment

This factor concerns systems exposed to marine atmospheres, moisture, acid rain, tropical temperature conditions, industrial chemicals, soils and dust in the atmosphere. Whenever possible, limit operation of aircraft in adverse environments.

Moisture is present in the air as a gas (water vapor) or as finely divided droplets of liquid (mist or fog) and often contains contaminants such as chlorides, sulfates and nitrates, which increase its corrosive effects. Condensed moisture which evaporates will leave contaminants behind. Condensed moisture and its contaminants can also be trapped in close fitting joints that may get wet such as faying surfaces can be drawn along poor bond lines by capillary action.

When dissolved in water, salt particles form strong electrolytes. Normal sea winds carry dissolved salt which makes coastal environments highly corrosive.

Industrial pollutants such as carbon, nitrates, ozone, sulfur dioxide, and sulfates contribute to the deterioration of nonmetallic materials and can cause severe corrosion of metals.

Warm moist air normally found in tropical climates accelerates corrosion while cold dry air normally found in arctic climates reduces corrosion.

Chemical Corrosion

Chemical corrosion occurs when a strong acid or base chemically dissolves metal. An example would be a piece of copper or zinc immersed in acid. The acid attacks and dissolves the metal until nothing is left. Aircraft are exposed to a number of corrosive chemicals. For example, aircraft batteries are a source of both strong acids and bases. In addition, acids can be produced in aircraft structures when certain conditions are present. It is common to find growths of microorganisms in aircraft fuel tanks. Certain bacteria have metabolisms that cause them to secrete corrosive fluids. These fluids form acids that attack metal structures.

Electrochemical Corrosion

Electrochemical corrosion occurs in the presence of metals that have various susceptibilities to corrosion. The mechanism involved is similar to the way a primary cell works. As you recall from your study of electricity, a primary cell consists of an anode, or donor of electrons, a cathode, or receiver of electrons and a conductive electrolyte. When a conductor completes the circuit between the anode and cathode, a chemical reaction occurs in the cell that produces electricity. As this process takes place, the anode is slowly worn away by the chemical reaction. The same process occurs in electrochemical corrosion.

A metal's electrode potential refers to the metal's ability to give up or receive electrons. Electrode potential is a function of a material's atomic structure; those with a high electric potential are strong electron donors while those with a low potential are poor electron donors. Table 2-1-1 lists those metals.

Because of its atomic structure, gold is not a good electron donor. Since it is not very susceptible to corrosion, gold is frequently used as a coating on electrical system contacts. At the other end of the scale is magnesium, which is a strong electron donor. This property makes magnesium extremely susceptible to corrosion. However, its strength and light weight make it indispensable for certain aircraft applications.

Electrochemical corrosion takes place when two different metals come in contact, either directly or through a conductive electrolyte. The further apart they are in their electrode potentials, the greater the degree of corrosion.

Moisture and oxygen are always present in the atmosphere. When metal is exposed to moisture, electrochemical corrosion has a place to start. One part of the metal surface will serve as a cathode, receiving electrons. Another part of the metal acts as an anode, giving up elec-

Group I	Magnesium and Magnesium Alloys
Group II	Aluminum, Aluminum Alloys, Zinc, Cadium and Cadium-Titanium Plate
Group III	Iron, Steels - Except Stainless Steels; Lead, Tin and their Alloys
Group IV	Copper, Brass, Bronze, Copper-Berylium, Copper-Nickel Chromium, Nickel, Nickel Base Alloys, Cobalt Base Alloys, Carbon Graphite, Stainless Steels, Titanium and Titanium Alloys

NOTE: 1. Metals listed in the same group are considered similar to one another.
2. Metals listed in different groups are considered dissimilar to one another.

Table 2-1-2. Metals safe to use in contact with one another.

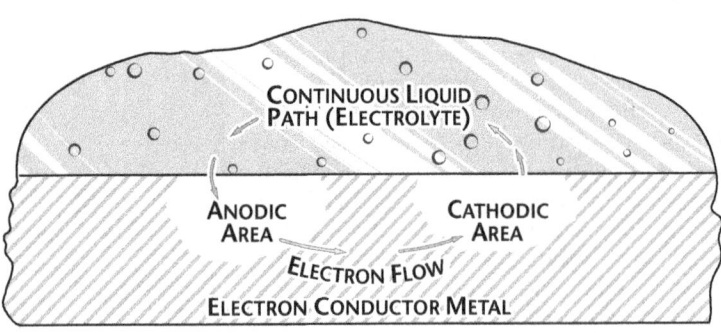

Figure 2-1-3. Water acts as an electrolyte when corrosion occurs on a metal structure

trons. Corrosion will always take place at the anode, as shown in Figure 2-1-3.

Several factors affect the spread of corrosion. If an aircraft is frequently operated near the ocean, the salt in the atmosphere increases the electrolyte action of the water and the corrosive process is more rapid. The same is true for air pollution. Sulfur and nitrogen compounds in the air dissolve in water and form acid compounds. These compounds attack structural components directly and provide a good electrolyte for electrochemical corrosion.

Section 2
Forms of Corrosion

The way corrosion attacks aircraft structures depends on a component's composition and location. The different types of corrosion are:

- Pitting corrosion starts on the surface and extends down into the material.
- Concentration cell corrosion occurs at joints between metal parts. The joint can be a butt splice, a seam, or an assembly.
- Oxygen concentration cells occur when the solution in contact with the metal surface contains dissolved oxygen. Figure 2-2-1 shows a bolt that is no longer usable.
- Filiform corrosion is a type of oxygen concentration cell corrosion that occurs on metal surfaces having an organic coating.
- Intergranular corrosion is an attack on the grain boundaries of a metal.
- Exfoliation corrosion is an advanced form of intergranular corrosion characterized by the lifting up of a metal surface's grains by the force of expanding corrosion products.
- Galvanic corrosion, one of the most common types of corrosion occurs when two dissimilar metals make contact in the presence of an electrolyte.
- Stress corrosion takes place when a metal part is subjected to a tensile load in a corrosive environment.
- Fatigue corrosion involves the cyclic stress of a part in a corrosive environment.
- Fretting corrosion happens when there is movement between two highly loaded surfaces that are not supposed to move.

Corrosive Agents

Corrosive agents are substances that cause a corrosive chemical reaction on metals. The most common corrosive agents are acids, alkalis (bases) and salts. The atmosphere and water, the two most common media for these agents, can sometimes act as corrosive agents themselves.

Acid. An acid is a chemical compound that gives up hydrogen ions when dissolved in water. Depending on the acid's strength, it can corrode most alloys used in aircraft structures. The most destructive are sulfuric acid, halogen acids (hydrochloric, hydrofluoric and hydrobromic) and organic acids found in human and animal waste.

Alkalis. Alkalis, or bases, are chemical compounds that give up hydroxyl ions when dissolved in water. Aluminum is vulnerable to corrosive attack from lime, washing soda, lye, and potassium hydroxide, an electrolyte used in nickel-cadmium batteries. Aluminum and magnesium alloys are generally more resistant to alkali attack than to acid attack.

Salts. Salts are chemical compounds formed by the chemical reaction that occurs when acids are mixed with alkalis. Sodium chloride, or table salt, is a good example. Solutions containing salts are good electrolytes that actively promote corrosive attack on aluminum and magnesium alloys. Many stainless steel alloys

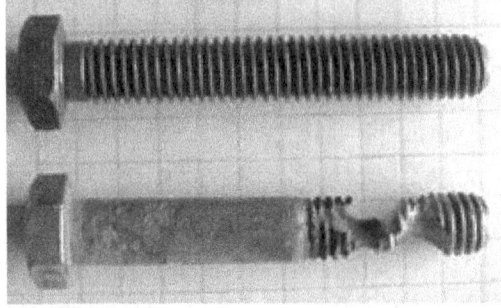

Figure 2-2-1. It is obvious how dangerous this condition can become

are resistant to attack by salt solutions. Aircraft operated in saltwater environments are especially vulnerable to salt-induced corrosion.

Mercury. Mercury is a heavy metallic element that is liquid at room temperature and is highly corrosive to aluminum alloys, stainless steel and brass. Through a process known as amalgamation, mercury chemically combines with these metals and produces severe corrosion. It can come in contact with aircraft structure through breaks in paint or other protective coatings, and when this happens the chemical attack is extremely rapid. Contamination results in pitting and intergranular attack, leaving the metal embrittled and weakened. Mercury and mercury compounds are frequently shipped on aircraft, and if spills occur, cleanup must be fast and thorough.

Water. Water acts as a corrosive agent on aircraft structures. The degree of corrosivity depends on the type and quantity of dissolved minerals, organic impurities and gases in the water.

The most corrosive natural waters are those that contain salts. Water in the open ocean is extremely corrosive, but waters in harbors are often more so because they are usually contaminated by industrial waste. The corrosiveness of fresh water varies depending on the kinds of dissolved impurities it contains. Certain industrial pollutants can make fresh water extremely corrosive.

Oxygen. The atmosphere contains oxygen, which when mixed with moisture in the air, acts as a corrosive agent. As with water, the presence of industrial pollutants greatly enhances the corrosiveness of the atmosphere around cities and industrial centers.

Organic growth. Organic growth includes bacteria and fungi that live in aircraft structures such as fuel tanks, water systems and galleys. They promote corrosion in several ways. Some microorganisms produce corrosive agents as waste products, which attack metal structures directly. Bacterial growths can entrap water around the floor structures of galleys, hastening the spread of corrosion.

Section 3
Corrosion Inspection

The importance of early corrosion detection cannot be overstated. If corrosion is found and repaired at an early stage of its development, there is less damage to the aircraft, the repair is less expensive to complete, downtime is reduced and safety is not compromised. The longer corrosion is left untreated, the bigger the problem it creates.

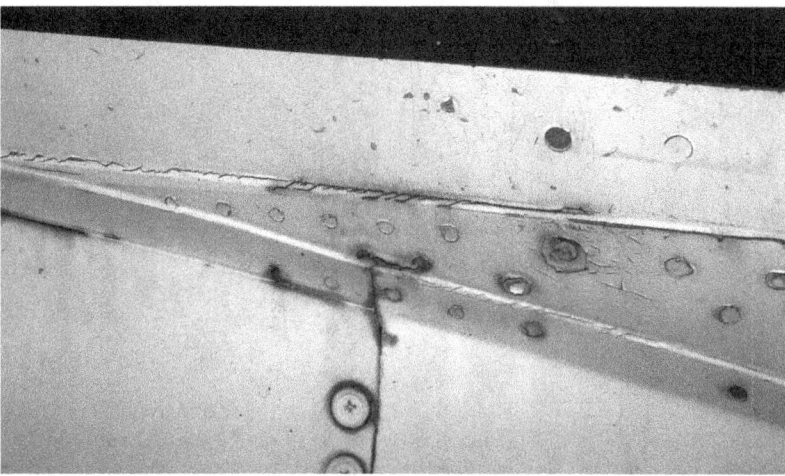
Figure 2-3-1. Cracks and pinholes expose metal to water and oxygen.

Dye penetrant, eddy current, ultrasonic and radiographic inspections are very effective in detecting corrosion damage. These methods are discussed in the section on nondestructive testing.

Visual inspection can be performed using a flashlight, magnifying glass and inspection mirror. On painted surfaces, the appearance and integrity of the paint is your best indicator of the condition of the metal beneath. Corrosion can change the color of the paint or cause the paint to have a scaly or blistered surface. Look for blisters or paint that is chipping and flaking off the surface. Watch for damage to the paint or to adjacent sealant. Figure 2-3-1 shows what can happen when cracks or pinholes in these areas expose the metal underneath to water and oxygen, the first step in the corrosive process.

On bare metal surfaces corrosion will appear as a dulled or darkened area with a pitted surface. You may also observe white, gray or reddish dust or particles.

Inspecting Corrosion-prone Areas

Some areas of an aircraft structure receive little exposure to corrosive conditions and thus experience less corrosion damage. Other areas, however, are constantly exposed to corrosives and need much more frequent inspection.

On both turbine and reciprocating engines, the high heat and corrosive compounds of engine exhaust can cause problems for exhaust components and structures in the exhaust gas path. Pay particular attention to gaps, seams, hinges and fairings in the exhaust gas path where deposits may be trapped and not reached by normal cleaning methods as shown in Figure

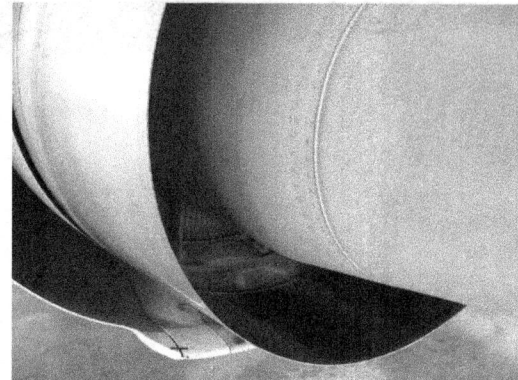

Figure 2-3-2. Exposure to high heat and corrosive exhaust gases make engine exhaust areas susceptible to corrosion.

Figure 2-3-3. Galley areas and lavatories are extremely corrosion-prone.

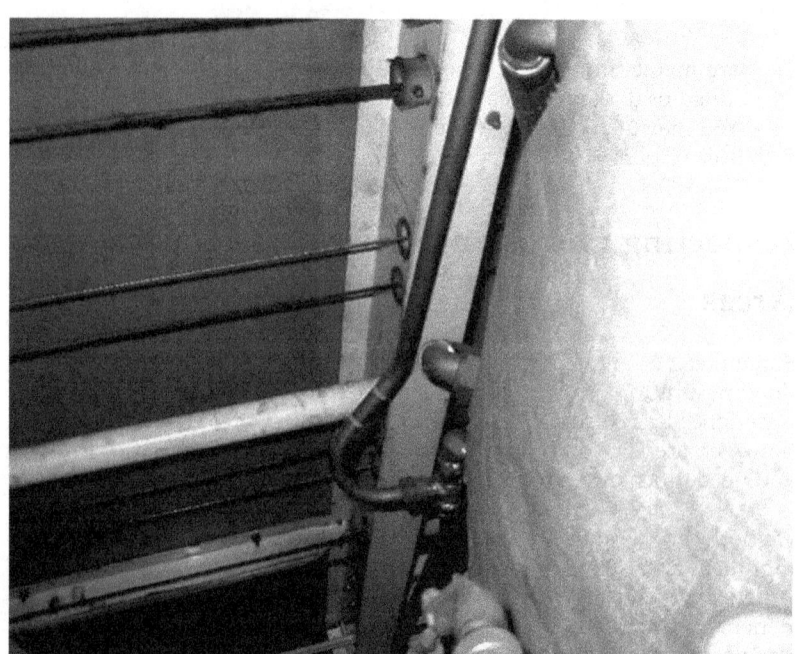

Figure 2-3-4. The area below the floorboards around galleys and lavatories is a likely site for corrosion. The large object on the right is a potable water tank, with a main floor beam running vertically in this picture.

2-3-2. This includes remote areas such as empennage structures.

Aircraft galley (see Figure 2-3-3) and lavatory areas are some of the most corrosion-prone areas you will encounter.

Deck areas behind lavatories, sinks and ranges where spilled food and waste products may collect are potential trouble spots. Even if some contaminants are not corrosive in themselves, they will attract and retain moisture and in turn cause corrosive attack. Figure 2-3-4 shows a potentially troublesome area.

Carefully inspect bilge areas located under galleys and lavatories, clean these areas frequently and keep paint touched up.

Aircraft battery electrolytes contain acid or strong alkali. As a result, battery compartments like those in Figure 2-3-5 and battery vent openings are frequently attacked by corrosion. Despite improvements in protective paint finishes and in methods of sealing and venting, battery compartments continue to be corrosion problem areas. Fumes from overheated electrolyte are difficult to contain and will spread to adjacent cavities and cause a rapid, corrosive attack on all unprotected metal surfaces.

Battery vent openings on the aircraft skin should be included in the battery compartment inspection and maintenance procedure. Regular cleaning and neutralization of acid deposits will minimize corrosion from this cause.

Landing gears and wheel wells probably receive more punishment than any other area on the aircraft. Mud, water, salt, gravel and other flying debris is picked up from ramps, taxiways or runways and thrown by tires. Because of the many complicated shapes, assemblies and fittings found in the wheel well and landing gear areas, complete area paint film coverage is difficult to attain. A partially applied preservative tends to mask corrosion rather than prevent it. Due to heat generated by braking action, preservatives cannot be used on some main landing gear wheels. During inspection of this area, pay particular attention to the following trouble spots:

- Magnesium wheels, especially around bolt heads, lugs and wheel-web areas, for the presence of entrapped water or its effects.

- Exposed rigid tubing, especially at B-nuts and ferrules, under clamps and tubing identification tapes.

- Exposed position indicator switches and other electrical equipment.

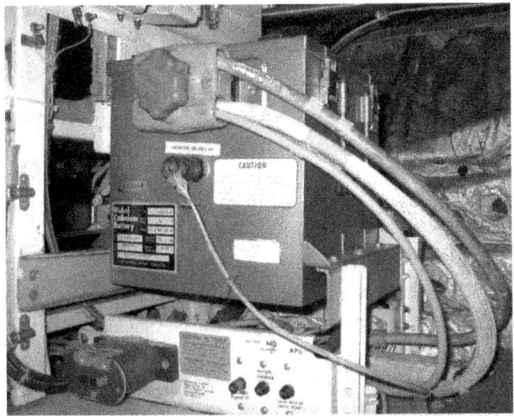

Figure 2-3-5. Battery compartments are breeding grounds for corrosion.

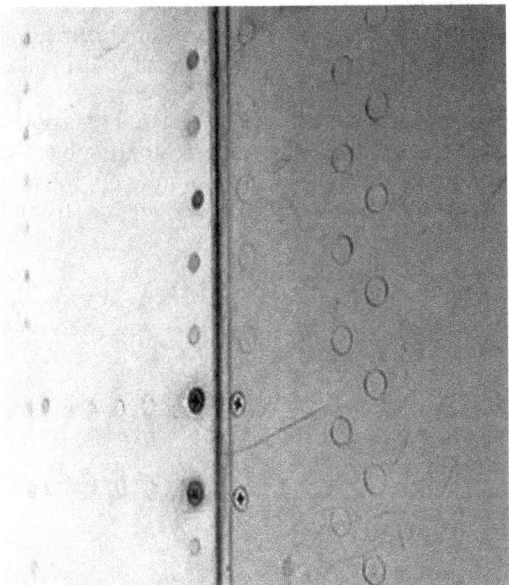

Figure 2-3-6. The two steel screws in this wing leading edge show signs of dissimilar metal corrosion.

- Crevices between stiffeners, ribs and lower skin surfaces, which are typical water and debris traps.

External aircraft surfaces are readily visible and accessible for inspection and maintenance. Even here, certain types of configurations or combinations of materials can become troublesome and require special attention. One example is shown in Figure 2-3-6.

Relatively little corrosion trouble is experienced with magnesium skins if the original surface finish and insulation are adequately maintained. Trimming, drilling, or riveting destroys some of the original surface treatment, which is never completely restored by touch-up procedures. Any inspection for corrosion should include all magnesium skin surfaces with special attention to edges, areas around fasteners and cracked, chipped, or missing paint.

Corrosion of metal skin joined by spot welding is the result of the entrance and entrapment of corrosive agents between the layers of metal. This type of corrosion is evidenced by corrosion products appearing at the crevices through which the corrosive agents enter. More advanced corrosive attack causes skin buckling and eventual spot-weld fracture. Skin buckling in its early stages may be detected by sighting along spot-welded seams or by using a straightedge. The only technique for preventing this condition is to keep potential moisture entry points such as seams and holes created by broken spot welds filled with a sealant or a suitable preservative compound.

Inspecting Inaccessible Areas

Fuel tanks. Because fuel tanks are usually located inside wing and fuselage structures, it is often difficult to gain access to inspect fittings and other hardware on the outside of the tank for corrosion. In addition fuel tanks are targets for bacterial growth, particularly in tanks used for turbine fuels. While inspection inside the tank may be difficult or impossible, bacterial growth can be controlled with the use of growth-inhibiting additives added to the fuel when refueling.

Piano hinges. Piano hinges are prime spots for corrosion due to the dissimilar metal contact between the steel pin and aluminum hinge. As you can see from Figure 2-3-7 they are also natural traps for dirt, salt, or moisture. Inspection of hinges should include lubrication and actuation through several cycles to ensure complete lubricant penetration.

Wing flap and spoiler recesses. These areas accumulate grease, dirt and water. Frequently they go unnoticed because flaps and spoilers are normally retracted. For this reason, these recesses are potential corrosion problem areas. Figure 2-3-8 illustrates the difficulty in inspecting these areas.

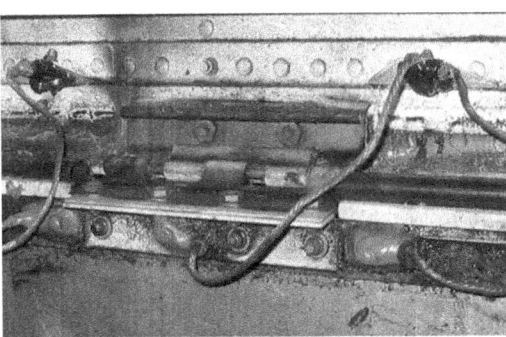

Figure 2-3-7. This piano hinge secures the air conditioning bay door to the fuselage on a Boeing 737. The hinge is aluminum, the pin steel. It is common to find these hinges corroded.

Figure 2-3-8. Flap recesses at wing trailing edge

Engine mounts. Because of their purpose and location, engine-mount structures are subjected to extremes of heat, vibration and torque from the engine and its accessories. To withstand the stresses placed on mount structures, most reciprocating engine mounts are manufactured from welded tubular steel. Therefore, engine-mount structures are inspected for corrosion in much the same manner as other tubular steel airframe components, such as push/pull tubes, airframe structural tubing, tubular landing gear, etc.

Particular attention must be paid to areas where moisture or other contaminants could possibly get inside the tubing, such as threaded, riveted, or welded areas. When economically feasible, corroded tubing should be cleaned, the structural integrity of the material tested (through the use of magnaflux, radiography, or other suitable test procedure) and treated to prevent a recurrence of the same or similar corrosion. When the cost is prohibitive, the alternative is replacement of the part. In some cases parts of the engine-mount structure can be individually replaced. In other cases the entire mount assembly must be replaced.

Control cables. All control cables, whether plain carbon steel or corrosion-resistant steel, should be inspected to determine their condition at each inspection period. Cables should be inspected for corrosion by random cleaning of short sections with solvent-soaked cloths. Control cable access is easier in large airplanes as shown in Figure 2-3-9 than in some smaller airplanes. If external corrosion is evident, tension should be relieved and the cable checked for internal corrosion. Cables with internal corrosion should be replaced. Light external corrosion should be removed with a stainless steel wire brush. When corrosion products have been removed, recoat the cable with preservative.

Inspecting Other Areas

Many types of fluxes used in brazing, soldering and welding are corrosive and will chemically attack the metals or alloys on which they are used. Therefore, it is important that residual flux be removed from the metal surface immediately after the joining operation. Flux residues are hydroscopic in nature; that is, they are capable of absorbing moisture, and unless carefully removed, tend to cause severe pitting.

Weld decay is a form of intergranular corrosion that attacks welds in stainless steel. It occurs because the process of welding often produces an undesirable heat treatment adjacent to the welded area, in turn producing separate phases of the metal, one of which may be preferentially attacked under adverse environmental conditions. Figure 2-3-10 shows a typical weld.

Electronic and electrical compartments cooled by ram air or compressor bleed air are subjected to the same conditions common to engine and accessory cooling vents and engine frontal

Figure 2-3-9. Most cable runs in large airplanes are accessible once the floorboards are removed.

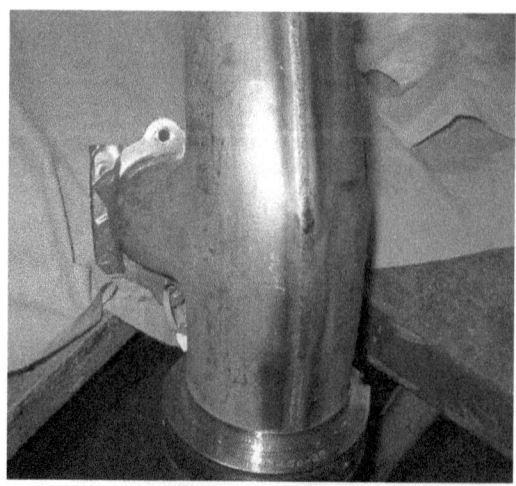

Figure 2-3-10. Welding often produces an undesirable heat treatment adjacent to the weld area.

areas. The degree of exposure is less because of a lower volume of air passing through as well special design features which to prevent water formation in the enclosed spaces. This is still a trouble area that requires special attention.

Circuit breakers, contact points and switches are extremely sensitive to moisture and corrosive attack and should be inspected for these conditions as thoroughly as design permits. If design features hinder examination of these items while in the installed condition, they should be carefully inspected when component is removed for other reasons.

Section 4
Corrosion Inspection Procedures for Aircraft

This section provides a general inspection guide for those parts or surfaces that can be visually inspected without complex disassembly of the aircraft. It is intended for use in establishing corrosion inspection areas for which the manufacturer has not provide a recommended corrosion inspection program. The manufacturer's recommended corrosion inspection program will take precedence over these guidelines. These inspections should be accomplished in conjunction with other preventive maintenance. Corrosion prone areas should be cleaned, inspected, and treated more frequently than less corrosion prone areas. The list is not necessarily complete for any specific aircraft, but could be used to set up a maintenance inspection program.

Exhaust Trail Areas

Both jet and reciprocating engine exhaust gas deposits are very corrosive. Inspection and maintenance of exhaust trail areas should include attention to these areas :

- Visually inspect paint in areas of the exhaust trails for damage.
- Visually inspect under fairings around rivet heads, and in skin crevices, for corrosion in areas of engine exhaust trail.
- Gaps, seams hinges and fairings are some of the exhaust trail areas where deposits may be trapped and not reached by normal cleaning methods. Exhaust deposit buildup on the upper and lower wing, aft fuselage, and in the horizontal tail surfaces will be considerably slower and sometimes completely absent from certain aircraft models.
- Inspection should also include the removal of fairings and access plates located in the exhaust gas path.

Battery Compartments and Battery Vent Openings

In spite of protective paint systems and extensive sealing and venting provisions, battery compartments continue to be corrosion problem areas, as can be seen in Figure 2-4-1. Unprotected surfaces will be subjected to corrosive attack. For lead-acid batteries, frequent cleaning and neutralization of acid deposits with sodium bicarbonate solution will minimize corrosion.

If aircraft batteries with electrolytes of either sulfuric acid or potassium hydroxide are in use, their leakage will cause corrosion. Consult the applicable maintenance manuals for the particular aircraft to determine which type of battery is installed and the recommended maintenance practices for each. Cleaning of nickel cadmium compartments should be done with ammonia or a boric acid solution, allowed to dry thoroughly, and then painted with an alkali-resistant varnish.

- Inspect the battery compartment for electrolyte spillage, corrosion, and condition of protective paint.
- Inspect the area around the battery vent for corrosion.

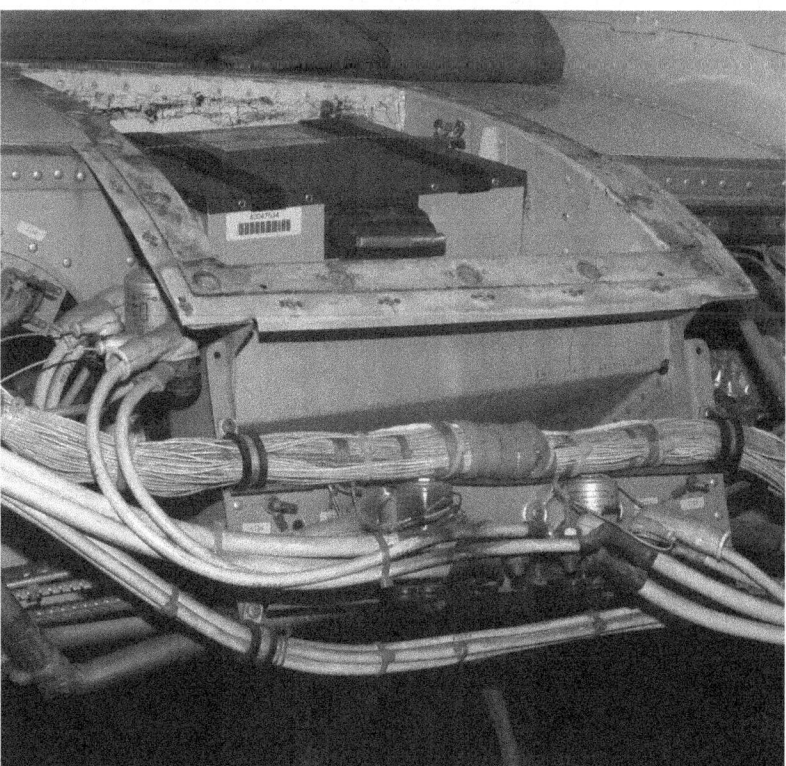

Figure 2-4-1. Battery compartments are problem areas.

Figure 2-4-2. Food preparation areas are important corrosion inspection items.

Figure 2-4-3. The under-floor area requires careful inspection and cleaning.

Lavatories and Galleys

Inspect areas around lavatories, sinks, and ranges for spillage and corrosion. Figure 2-4-2 is an example of an aircraft galley. Pay particular attention to the floor area and the area behind and under lavatories, sinks and food heating units where spilled food and waste products may collect. If these areas are not kept clean they become potential trouble spots.

Bilge Areas

On all aircraft, the bilge area is a common trouble spot, as shown in Figure 2-4-3. The bilge is a natural sump or collection point for waste hydraulic fluids, water, dirt, loose fasteners, drill chips, and other odds and ends of debris.

Residual oil can mask small quantities of water which settle to the bottom and set up a hidden potential corrosion cell. With the exception of water displacing corrosion preventative compounds, keeping bilge areas free of all extraneous material including water and oil will insure the best protection against corrosion.

- Inspect bilge areas for waste hydraulic fluids, water, dirt, loose fasteners, drill chips, and other debris.
- Remove any foreign material from bilge and inspect for corrosion.

Wheel Wells and Landing Gear

The wheel well area probably receives more punishment than any other area of the aircraft. It is exposed to mud, water, salt, gravel, and other flying debris from runways during flight operations.

Inspect wheel well area and landing gear components for damage to exterior finish coating and corrosion.

Frequent cleaning lubrication and paint touch-up are needed on aircraft wheels and on wheel well areas, as shown in Figure 2-4-4. Because of the many complicated shapes, assemblies, and fittings in the area, complete coverage with a protective paint film is difficult to attain. Thus, preservative coatings tend to mask trouble rather than prevent it. Because of the heat generated from braking, preservative coatings cannot be used on aircraft landing gear wheels.

During inspection of wheel wells and landing gear particular attention should be given to the following trouble spots:

- High strength steel
- Exposed surfaces of struts, oleos, arms, links, and attaching hardware (bolts, pins, etc.)
- Axle interiors
- Exposed position indicator switches and other electrical equipment
- Crevices between stiffeners, ribs, and lower skin surfaces which are typical water and debris traps
- Magnesium wheels, particularly around bolt heads, lugs, and wheel web areas
- Exposed rigid tubing, especially at B-nuts and ferrules under clamps and tubing identification tapes

External Skin Areas

External aircraft surfaces are ordinarily covered with protective finishes. In addition, paint coatings may be applied. The affected external aircraft surfaces are readily visible or available for inspection and maintenance. Much emphasis

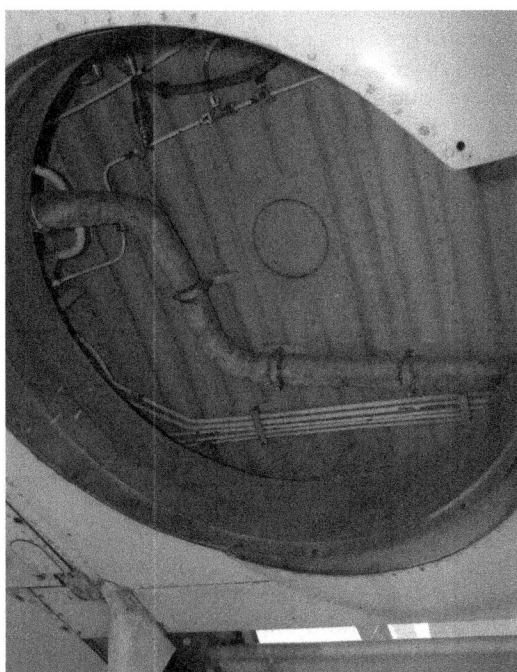

Figure 2-4-4. Wheel wells are traps for dirt and water.

Faying Surfaces and Crevices

Corrosion in faying surfaces, seams, and moisture and other corrosive agents. They are usually detectable by bulging of the skin. See Figure 2-4-6.

Magnesium Skins. Properly surface treated, insulated and painted magnesium skin surfaces are rarely problematic from a corrosion standpoint if the original surface is maintained. However, trimming, drilling, and riveting destroy some of the original surface treatment which may not be completely restored by touch-up procedures.

Some aircraft have steel fasteners installed through magnesium skin with only protective finishes under the fastener heads, and fillet sealant or tape over the surface for insulation. Further, all paint coatings are inherently thin at abrupt changes in contour, such as at trimmed edges. With magnesium's sensitivity to moisture, all of these conditions add up to a potential corrosion problem whenever magnesium is used.

has been given to these areas in the past, and maintenance procedures are well established. Even here, certain types of configurations or combinations of materials become troublesome and require special attention if serious corrosion difficulties are to be avoided. This is shown in Figure 2-4-5. Some of the common trouble areas, other than those attributed to engine exhaust deposits, are grouped as follows:

- External skin surfaces for damage to protective finishes and corrosion.
- The area around fasteners for damage to protective finishes and corrosion.
- Lap joints for bulging of skin surface, which may indicate the presence of corrosion between the faying surfaces. Skin cracks and/or dished or missing fastener heads may also indicate severe corrosion in bonded joints.
- The area around spot welds for bulges, cracks, or corrosion.
- Thick alloy skin surfaces for pitting, intergranular corrosion, and exfoliation of the metal. Look for white or gray deposits around countersunk fastener heads and raised areas or bumps under the paint film.
- Composite skins for corrosion of attachment fasteners.
- Steel titanium CRES and nickel alloy fasteners and areas around these fasteners are trouble spots. These areas are subject to high operational loads, moisture intrusion, and dissimilar metal skin corrosion.

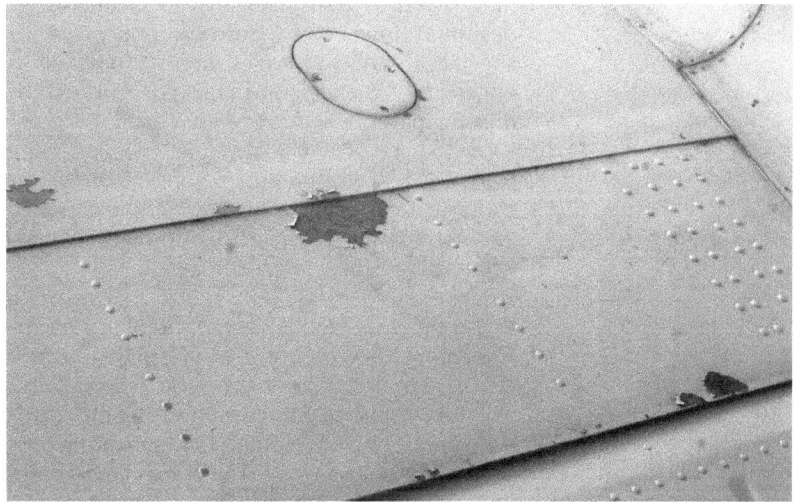

Figure 2-4-5. The external skin is subject to weathering of the paint.

Figure 2-4-6. All faying surfaces or lap joints must be inspected.

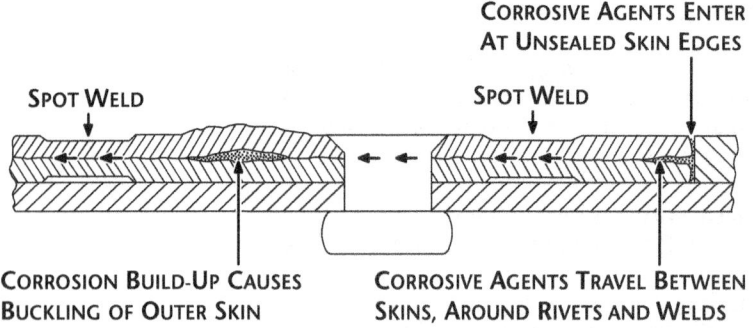

Figure 2-4-7. Bulging skin is caused by corrosion byproducts.

Figure 2-4-8. This hinge is starting to show evidence of fretting corrosion.

Any inspection for corrosion should include all magnesium skin surfaces, as well as other magnesium fittings or structural components, with special attention to edges, areas around skin edges and fasteners, and cracked, chipped, or missing paint.

Spotwelded skins. Corrosion of this type construction is chiefly the result of the entrance and entrapment of moisture or other corrosive agents between layers of the metal.

Spotwelded assemblies are particularly corrosion prone. See Figure 2-4-7. Corrosive attack causes skin buckling or spotweld bulging, and eventually spotweld fracture. Some of the corrosion may be caused originally by fabricating processes, but its progress to the point of skin bulging and spotweld fracture is the direct result of moisture or other corrosive agents working its way through open gaps or seams. The use of weld-through sealing materials is expected to minimize this problem, but many in-service aircraft still have unsealed spotweld skin installed. This type of corrosion is evidenced by corrosion products appearing at the crevices through which the corrosive agents enter.

Corrosion may appear at either external or internal faying surfaces, but it is usually more prevalent on external areas. More advanced corrosive attack causes skin buckling and eventual spotweld fracture. Skin buckling in this early stage may be detected by sighting or feeling along spotwelded seams or by using a straight edge.

To prevent this condition, keep potential moisture entry points including gaps, seams, and holes created by broken welds filled with noncorrosive sealant.

Piano hinges. These are prime spots for corrosion due to dissimilar metal contact between the steel pin and aluminum hinge tangs. See Figure 2-4-8. They are also natural traps for dirt, salt, and moisture. Where this type of hinge is used on access doors or plates, and actuated only when opened during an inspection, they tend to corrode and freeze in the closed position between inspections. When the hinge is inspected, it should be lubricated and actuated through several cycles to ensure complete penetration of the lubricant. See Figure 2-4-9.

Inspect all bonding jumpers across the hinge for damage and security.

Heavy or tapered aluminum alloy skin surfaces. Heavy or thick sections of most heat-treated aluminum alloys are susceptible to pitting or intergranular corrosion and metal exfoliation. When inspecting external skin surfaces, especially around countersunk fastener heads, look for white or grey powder deposits or metal exfoliation. This first becomes evident as small raised areas or bumps under paint film.

Treatment of this type of corrosive attack includes removal of all corrosion products. Exfoliated metal is blended and polished not to exceed the limits set by the aircraft manufacturer. If corrosion products remain after the limits set by the aircraft manufacture have been reached, contact the aircraft manufacturer or the Federal Aviation Administration (FAA) for authorized limits. Treatment is not complete until the protective surface finishes are restored.

Protect reworked areas with a chemical conversion coating, sealant primer, and top coat if applicable. Reworked areas should be carefully watched for any indications of renewed corrosive activity.

Organic composites. Organic composites used in aircraft can cause different corrosion problems than those normally associated with

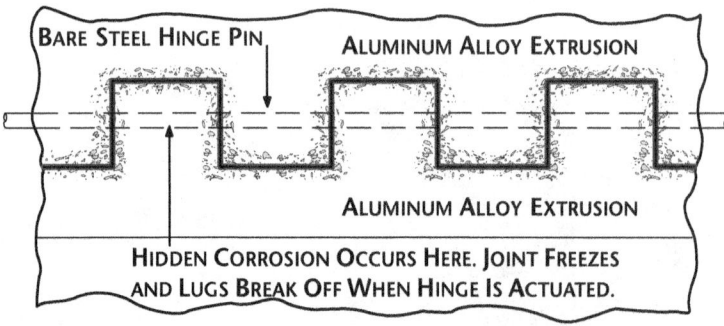

Figure 2-4-9. Piano hinges are examples of dissimilar metal corrosion.

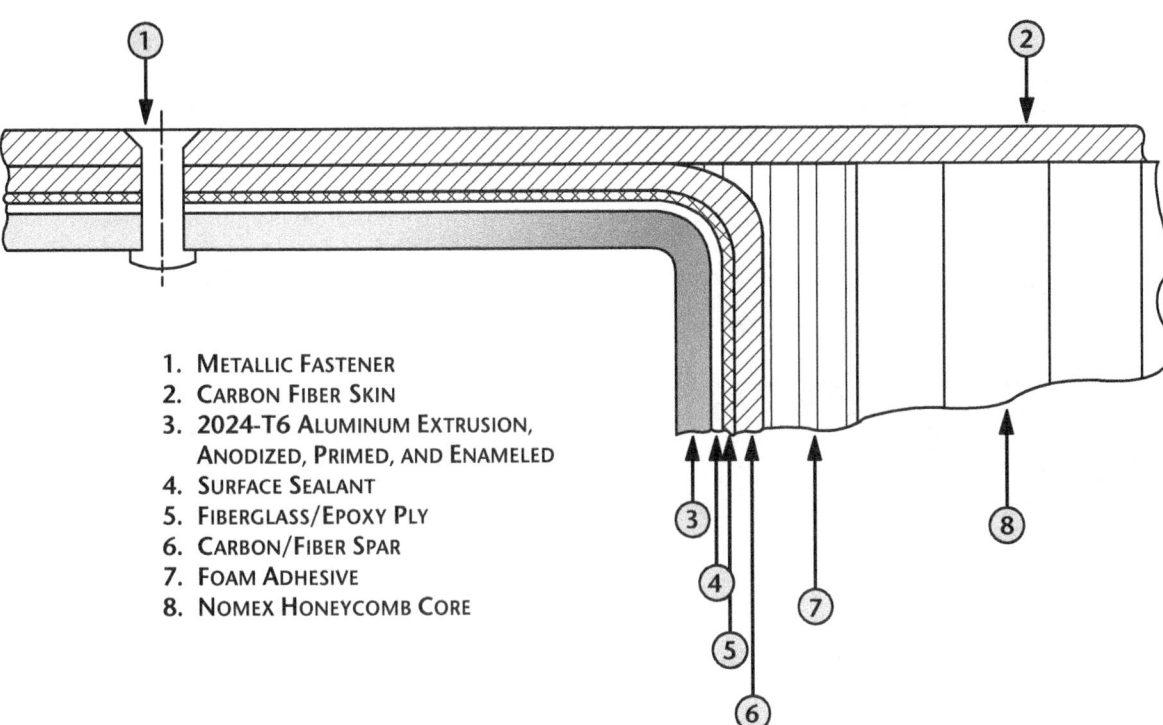

1. Metallic Fastener
2. Carbon Fiber Skin
3. 2024-T6 Aluminum Extrusion, Anodized, Primed, and Enameled
4. Surface Sealant
5. Fiberglass/Epoxy Ply
6. Carbon/Fiber Spar
7. Foam Adhesive
8. Nomex Honeycomb Core

Figure 2-4-10. Separating graphite from aluminum is important to prevent corrosion.

all metal structures. Composites such as graphite/epoxy act as a very noble (cathodic) material, creating the potential for galvanic corrosion. The galvanic corrosion potential coupled with different methods of attachment (i.e., adhesive bonding, stepped structures, locking mechanical fasteners, etc.) lead to multicomponent galvanic couples. The problem is particularly aggravated by high humidity and salt water environments.

Application of aircraft sealants over dissimilar metal/composite junctions will prevent galvanic corrosion if moisture is completely excluded. See Figure 2-4-10. However, since complete exclusion of moisture is virtually impossible under extended periods of flight operation, the most effective method of eliminating the voltage potential is to provide nonconductive layers such as fiberglass/epoxy and/or sealant between the composite and dissimilar metal surfaces.

Water Entrapment Areas

Corrosion will result from the entrapment of moisture. With the exception of sandwich structures, design specifications usually require that the aircraft have low point drains installed in all areas where moisture and other fluids can collect. In many cases, these drains are ineffective either due to location or because they are plugged by sealants, extraneous fasteners, dirt, grease, and debris. Potential entrapment areas are not a problem when properly located drains are functioning, and the aircraft is maintained

Figure 2-4-11. Engine frontal areas are subject to corrosion.

in a normal ground attitude. However, the plugging of a single drain hole or the altering of the level of the aircraft can result in a corrosion problem if water becomes entrapped in one of these "bathtub" areas. Daily inspection of low point drains is a recommended practice.

Engine frontal areas. Constant abrasion by airborne dirt and dust, bits of gravel from runways, and rain tends to remove the protective surfaces from these areas. Figure 2-4-11 is a good example of this area. Furthermore, cores of radiator coolers, reciprocating engine cylinder fins, etc., due to the requirement for heat dissipation, may not be painted. Engine acces-

Figure 2-4-12. Reciprocating engines and propellers should be carefully inspected.

sory mounting bases usually have small areas of unpainted magnesium or aluminum on the machined mounting surfaces. With moist and salt or industrial pollutant-laden air constantly flowing over these surfaces, they are prime sources of corrosive attack.

Inspection of such areas should include all sections in the cooling air path with special attention to obstructions and crevices where salt deposits may build up during marine operations. Figure 2-4-12 is a radial engine frontal area. Inspection procedures include:

- Inspect reciprocating engine cylinder fins, engine cases, and cooling air ducts for damage to exterior finish and corrosion.
- Inspect radiator cooler cores for corrosion.

Electronic Package Compartments

Electronic and electrical compartments cooled by ram air or compressor bleed air are subjected to the same conditions common to engine and accessory cooling vents and engine frontal areas. While the degree of exposure is less because of a lower volume of air passing through and special design features incorporated to prevent water formation in enclosed spaces, this is still a trouble area that requires special attention as can be seen in Figure 2-4-13.

If design features hinder the examination of these items while in the installed condition, inspection should be accomplished whenever components are removed for other reasons.

Most corrosion that occurs on avionics equipment is similar to that which occurs on the basic airframe structure. The difference between avionic and airframe corrosion is that minute amounts of corrosion in avionics equipment can cause serious degradation or complete failure, while it would be unnoticed on larger structures.

Smog smoke soot and other airborne contaminants are extremely corrosive to exposed avionic equipment. Many fumes and vapors emitted from factories or industrial complexes are highly acidic and greatly accelerate corrosion. An example is the corrosive effect of ozone, a product of many welding machines and large electrical motors. Complete degradation of rubber seals and damage to delicate components have occurred in equipment stored near ozone-producing equipment. Avionics shops and storage areas should have a filtered air-conditioning system.

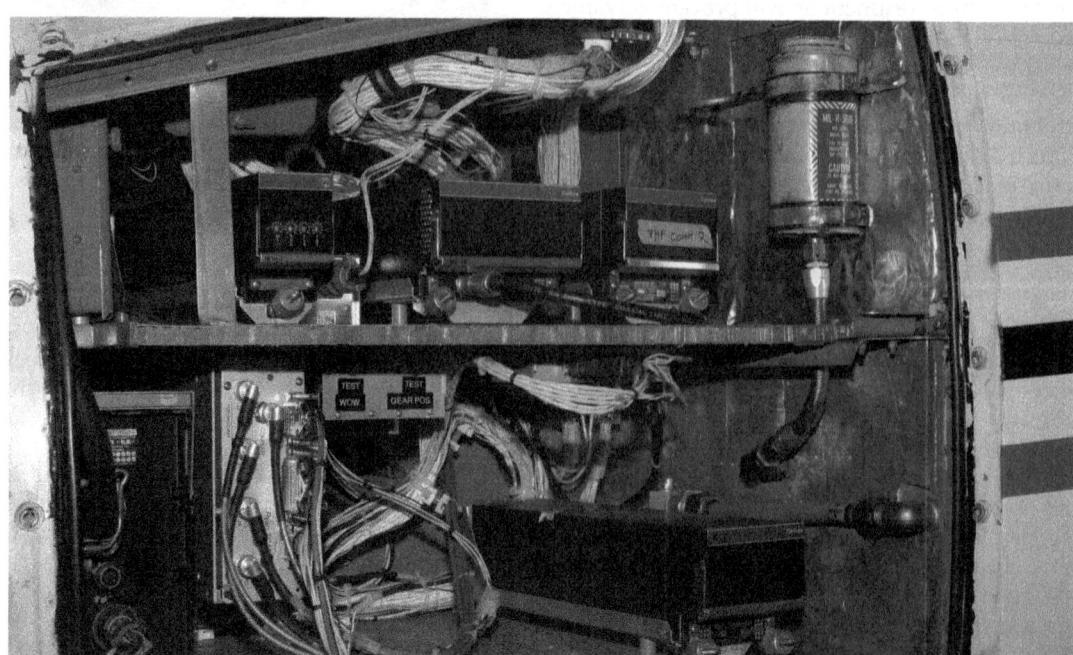

Figure 2-4-13. Electronics bays require special attention and inspection.

Another man-made atmosphere is the aircraft environmental control system. These systems induce cooling air to the equipment. They may include a filter system that extracts moisture, and in some cases contaminants, from the air that enters the equipment. Failure to replace and/or clean a filter, or eliminate a leaky environmental seal, may cause a moisture or contaminant buildup that could cause a corrosive atmosphere within the equipment.

Electrical connectors. Electrical connectors may be potted with a sealing compound to provide more reliability of equipment. Figure 2-4-14 shows multiple potted connectors. The sealing compound prevents entrance of moisture into the area of connectors where the wires are attached to the pins.

O-rings may also be used to seal moisture out of the mating area of pin connections and to prevent loss of pressurization in compartments containing bulkhead connectors.

Moisture intrusion into electrical connectors can cause corrosion and an electrical failure. Suspected plugs should be disconnected, dissembled, solvent cleaned, and inspected for corrosion.

When sealing provisions are not designed into the electrical component, these components can have moisture intrusion and internal corrosion. Common inspection procedures include:

- Inspect electrical connectors for breaks in potting compound and corrosion of pins and wires.
- If the electrical connector is suspected of having moisture intrusion, disassemble the connector, clean the connector, and inspect it for corrosion.

Other Inspection Areas

Pressure bulkheads. Severe corrosion damage to the rear pressure bulkhead below the floor may occur as a result of contamination by fluids. The difficulty of inspection is illustrated in Figure 2-4-15. Inspection for rear bulkhead corrosion may require extensive disassembly of components and fixtures to allow a thorough visual inspection.

When inspection access holes are available, inspection by fiber optics is useful. Other non-destructive inspection (NDI) methods (X-ray, ultrasonic, and eddy current) are also available. However, these inspection techniques require specially trained personnel, NDI comparison standards, and suitable access. A regular inspection of the rear pressure bulkhead (both front and rear faces) below the floor level should be accomplished to prevent serious corrosion from occurring between the bulkhead and periphery doubler at the floor level. Such corrosion could weaken the bulkhead skin and cause sudden cabin pressure loss.

Bonded joints. Some older aircraft have developed delaminations in cold bonded joints. Figure 2-4-16 is an example of these types of joints. Corrosion between the delaminated surfaces is caused by moisture intrusion along the edge of the mating parts or around fasteners securing the mating parts together. Localized bulging of the skin or internal structural com-

Figure 2-4-14. Electrical connectors must be inspected and cleaned properly.

Figure 2-4-15. The bottom sections of pressure bulkheads can be sources of corrosion.

2-16 | Corrosion Inspection

Figure 2-4-16. Bonded joints.

Figure 2-4-17. Flexible hoses need periodic inspections.

ponent, usually around the fasteners, is the first indication of a corrosion problem. Skin cracks or dished or missing fastener heads may also indicate severe corrosion in bonded joints.

Corrosion which occurs between skins, doublers, and stringers or frames will produce local bulging or pulled rivets. Corrosion that occurs between the skins and doublers or tear straps away from backup structure such as stringer or frame will not produce local bulging.

An external low frequency eddy current inspection may be used to determine the extent of corrosion in the skin. Lap joints should be opened with wedges to determine the full extent of corrosion damage. Internal visual inspection should be used to detect delaminated doublers or tear straps. A penetrating water displacement corrosion inhibitor should be applied to faying surfaces after corrosion removal and repair.

Flexible hose assemblies. Examine all flexible hose assemblies for chafing, weather checking, hardening, discoloration, evidence of fungus, and torn weather protective coatings or sleeves.

This can be seen in Figure 2-4-17. Replace any defective, damaged, twisted, or bulging hoses.

Sandwich panels. Inspect edges of sandwich panels for damage to the corrosion protection finish or sealant and for corrosion. See Figure 2-4-18.

Trimmed edges of sandwich panels and drilled holes should have some type of corrosion protection. A brush treatment with an inhibitor solution or the application of a sealant along the edge, or both, is recommended. Any gaps or cavities where moisture, dirt, or other foreign material can be trapped should be filled with a sealant. The adjacent structure (not the sandwich) should have sufficient drainage to prevent moisture accumulation. Damage or punctures in panels should be sealed as soon as possible to prevent additional moisture entry—even if permanent repair has to be delayed.

Control cables. Control cables may present a corrosion problem whether carbon steel or stainless steel is used. The difficulty can be seen in Figure 2-4-19. The presence of bare spots in the preservative coating is one of the main contributing factors in cable corrosion.

Cable condition should be determined by cleaning the cable assembly, inspection for corrosion, and application of an approved preservative if no corrosion is found. If external corrosion is found, relieve tension on the cable and check internal strands for corrosion. Cables with corrosion on internal strands should be replaced.

Pay particular attention to sections passing through fairleads, around sheaves, and grooved bellcrank arms. External corrosion should be removed by a clean, dry, coarse rag or fiber brush. After complete corrosion removal, apply a preservative.

Figure 2-4-18. The edges of sandwich panels should be sealed.

Integral fuel cells. Sealant materials (Buna N Polyurethane and Epoxy) used in integral fuel cells are impervious to fuel but not completely impervious to moisture absorption. Since it is impossible to keep fuel completely free of water, moisture can penetrate through the topcoating materials and sometimes cause pitting or intergranular corrosion on aircraft structural parts.

It has also been found that micro-organisms which live in the water entrained by fuel, particularly jet propellant types, feed on fuel hydrocarbon and hydrocarbon-type elastomeric coating materials. These micro-organisms excrete organic acids, and dead micro-organisms act as a gelatinous acidified sponge which can deteriorate integral tank coatings and corrode the aircraft structure.

Microbial corrosion can be minimized by preventing as much water contamination of the fuel as possible with well-managed storage facilities, adequate filtration of fuel, and drainage of water contamination from integral fuel cells which keeps the water moving and reduces the chance for the colonies of micro-organisms to develop. Micro-organic activity can be reduced by using a biocide additive. Solution strength and application frequency should be in accordance with the manufacturer's instructions.

Protective finishes. These protect the base material from corrosion and other forms of deterioration. Protective finishes are divided into sacrificial and non-sacrificial categories. Sacrificial coatings include cadmium, zinc, and aluminum. Non-sacrificial coatings include hard plating (chromium and nickel), chemical conversion coatings, sealant, primers, and top coat.

At a minimum, finishes should be inspected as follows:

- Inspect top coat finish for breaks, peeling, lifting of surface, or other damage.
- Inspect aircraft structure for top coat finish damage from pitting or intergranular corrosion.

Section 5

Inspection Requirements

Except for special requirements in trouble areas, inspection for corrosion should be a part of routine maintenance inspections; i.e. daily or preflight. Overemphasizing a particular corrosion problem when it is discovered and forgetting about corrosion until the next crisis is an unsafe, costly, and troublesome practice.

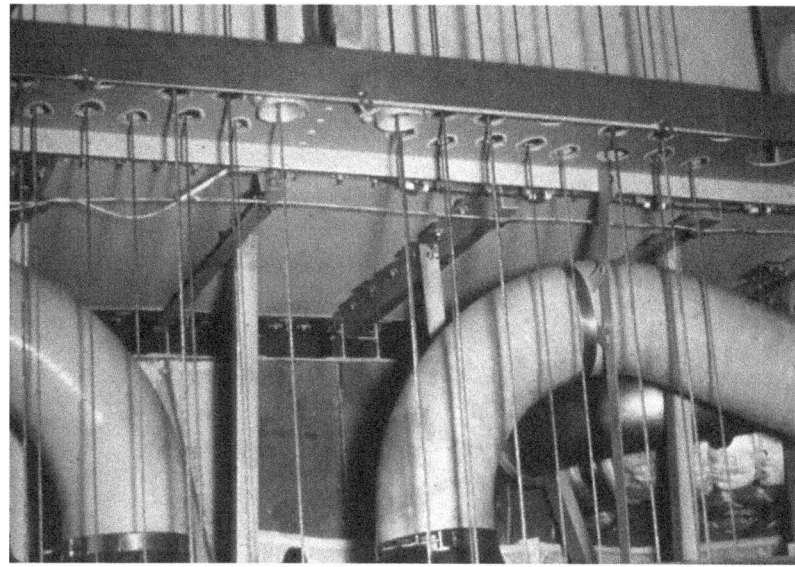

Figure 2-4-19. Control cables require careful inspection for corrosion.

If corrosion control is assigned to a special team or shift, maintenance checks should be scheduled in such a way that these teams may accomplish their inspections and necessary rework while access plates are removed and components are disconnected or out of the way. Figure 2-5-1 shows a crew member at work.

Most manufacturers handbooks of inspection requirements are complete enough to cover all parts of the aircraft or engine, and no part or area of the aircraft should go unchecked. Use these handbooks as a general guide when an area is to be inspected for corrosion.

Trouble areas are a different matter. Experience shows that certain combinations of conditions

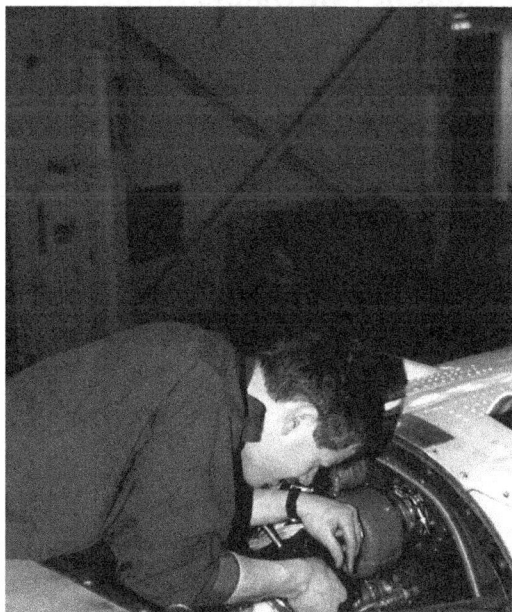

Figure 2-5-1. A corrosion crewmember inspecting part of an aircraft

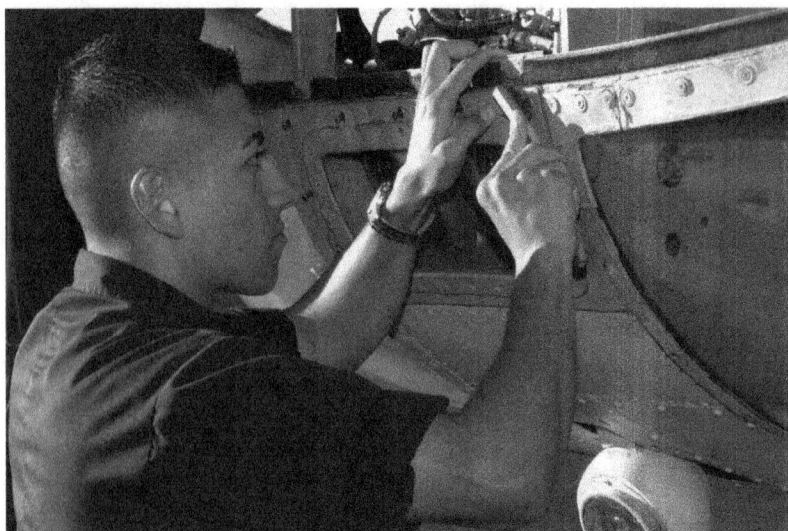
Figure 2-5-2. Remove preservative coatings for inspection.

result in corrosion in spite of routine inspection requirements. These trouble areas may be peculiar to particular aircraft models, but similar conditions are usually found on most aircraft.

The flight routes and bases of operation will expose some airplanes to more corrosive conditions than others. The operational environment of an aircraft may be categorized as mild, moderate, or severe, with respect to the corrosion severity of the operational environment. The corrosion severity of any particular geographic area may be increased by many factors including airborne industrial pollutants, chemicals used on runways and taxiways to prevent ice formation, humidity, temperatures, prevailing winds from corrosive environment, etc.

Frequency of Inspections

In addition to the routine maintenance inspections, the following special requirements should be observed:

- Aircraft operating in severe environments should be inspected every 15 days.
- Aircraft operating in moderate environments should be inspected every 45 days.
- Aircraft operating in mild environments should be inspected every 90 days.
- The aircraft should be washed prior to any inspection for corrosion.
- Checks should be performed by a crew familiar with corrosion problems and the nature of their treatment.
- Operators of low utilization aircraft should develop a corrosion inspection and repair program based on calendar time rather than flight hours.
- Due to the uncertainties that may be encountered in various operating environments, adjustments to the calendar time inspection interval should be made after analysis of corrosion inspection findings.

Recommended Depth of Inspection

The applicability of inspection requirements provide a means to insure adequate inspection of all compartments and interior aircraft cavities. When general requirements are observed, along with a periodic check of the list of common trouble areas, adequate maintenance is be assured for most operating conditions. To assist in assuring complete coverage, the following summary should be followed:

- Daily and preflight inspection.
- Check engine compartment gaps, seams, and faying surfaces in the exterior skin.
- Check all areas which do not require removal of fasteners, panels, etc., such as bilge areas, wheel and wheel well areas, battery compartments, fuel cell and cavity drains.
- Check engine frontal areas, including all intake vents, and engine exhaust areas.

In addition to the more common trouble spots that are readily available for inspection, remove screw-attached panels, access plates, and removable skin sections as necessary to thoroughly inspect the internal cavities. Inspection should also include removal of questionable heavy preservative coatings, at least on a spot-check basis. Figure 2-5-2 is an example of this area. Inspect the interior of the aircraft in corrosion prone areas such as around lavatories, galleys, under floors, baggage compartments, etc.

- Corrosion inspections should be accomplished at each annual inspection or other scheduled in-depth inspection in which areas of the aircraft not normally accessible will be available for corrosion inspection.
- Corrosion preventive compounds such as LPS3, Dinol AV5, or equivalent products and later advanced developments of such compounds may be used to effectively reduce the occurrence of corrosion. Results of corrosion inspections should be reviewed to help establish the effectiveness of corrosion preventive compounds and determine the reapplication interval.

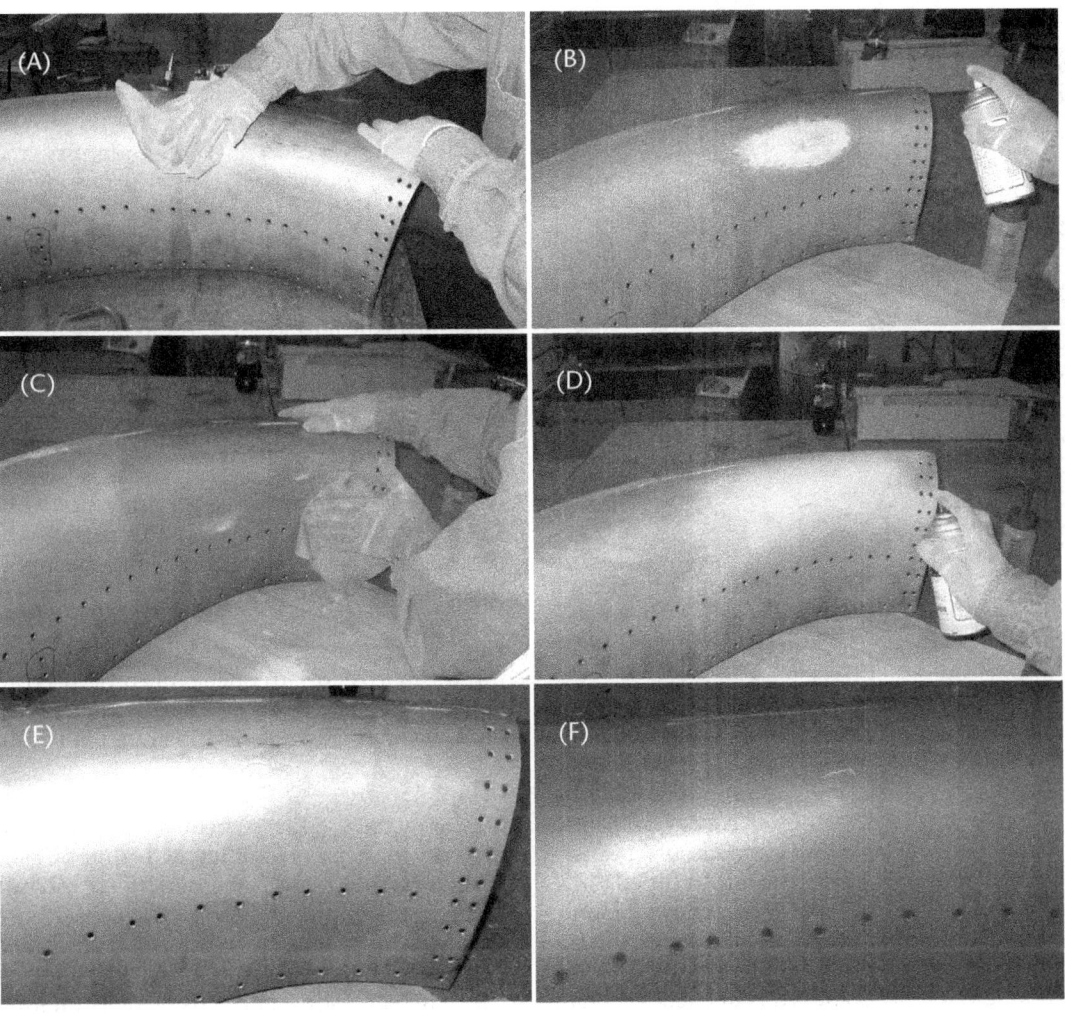

Figure 2-5-3.
(A) The technician cleans the area to be inspected.

(B) Apply the penetrant using a spray can.

(C) After allowing the correct time for penetration, all traces of the penetrant must be removed.

(D) Apply the developer using a spray can. Normally, the correct amount is about half as much as the penetrant.

(E) The damaged area after the excess developer has been removed

(F) The crack is revealed under ultraviolet light. This crack could not be repaired and the part was scrapped.

Primary Approach

The primary approach to corrosion detection is corrosion inspections on a regularly scheduled basis. Early detection and treatment reduces costs, out of service time, and the possibility of flight or flight related incidents. All corrosion inspections should start with a thorough cleaning of the area to be inspected. A general visual inspection of the area follows using a flashlight, inspection mirror, and a 5 to 10 power magnifying glass. The general inspection should look for obvious defects and suspected areas. A detailed inspection of damage or suspected areas found during the general inspection should follow.

Nondestructive Inspection (NDI)

Visual inspection. Visual inspection is the most widely used technique and is an effective method for the detection and evaluation of corrosion. Visual inspection employs the eyes to look directly at an aircraft surface, or at a low angle of incidence to detect corrosion. Using the sense of touch of the hand is also an effective inspection method for the detection of hidden well developed corrosion. Other tools used during the visual inspection are mirrors, borescopes, optical micrometers, and depth gauges.

The indications of corrosive attack can take several forms depending on the type of metal and the length of time the corrosion has had to develop. Corrosion deposits on aluminum and magnesium are generally a white powder, while ferrous metals vary from red to dark reddish brown stains.

Sometimes the inspection areas are obscured by structural members, equipment installations, or for some other reason are awkward to check visually. Remove access panels and adjacent equipment, cleaning the area as necessary, and remove loose or cracked sealants and paints. Mirrors, borescopes, and fiber optics are useful in observing obscure areas.

Methods of NDI Inspection

In addition to visual inspection there are several NDI methods such as liquid penetrant, magnetic particle, eddy current, X-ray, ultra-

Figure 2-5-4. Magnetic particle inspection for corrosion on a nose gear.

a blotter and draws the dye from the cracks or fissures back to the surface of the part, giving visible indication of the location of any fault that is present on the surface. The magnitude of the fault is indicated by the quantity and rate of dye brought back to the surface by the developer.

Magnetic particle inspection. This method may be used for the detection of cracks or flaws on or near the surface of ferromagnetic metals (metals which are attracted by magnetism). This inspection method is shown in Figure 2-5-4. A portion of the metal is magnetized, and finely divided magnetic particles (either in liquid suspension or dry) are applied to the object. Any surface faults will create discontinuities in the magnetic field and cause the particles to congregate on or above these imperfections, thus locating them.

Eddy current inspection. Eddy current testing (primarily low frequency) can be used to detect thinning due to corrosion and cracks in multi-layered structures is illustrated in Figure 2-5-5.

Low frequency eddy current testing can also be used to some degree for detecting or estimating corrosion on the hidden side of aircraft skins because when used with a reference standard the thickness of material which has not corroded can be measured. Low frequency eddy current testing can be used for estimating corrosion in underlying structure because the eddy currents will penetrate through into the second layer of material with sufficient sensitivity for approximate results.

High-frequency eddy current testing is most appropriate for detection of cracks which penetrate the surface of the structure on which the eddy current probe can be applied (including flat surfaces and holes).

X-ray inspection. X-ray inspection has somewhat limited use for the detection of corrosion because it is difficult to obtain the sensitivity

sonic, and acoustical emission which may be of value in the detection of corrosion. These methods have limitations and should be performed only by qualified and certified NDI personnel. Eddy current, X-ray, and ultrasonic inspection methods require that the equipment be calibrated each time it is used and a controlling reference standard is needed to obtain reliable results.

Liquid dye penetrant inspection. Inspection of large stress-corrosion or corrosion fatigue cracks on nonporous ferrous or nonferrous metals may be done by using of liquid dye penetrant processes. This process is shown in Figure 2-5-3. The dye applied to a clean metallic surface will enter small openings, such as cracks or fissures by capillary action. After the dye has had an opportunity to be absorbed by any surface discontinuities, the excess dye is removed and a developer is applied to the surface. The developer acts like

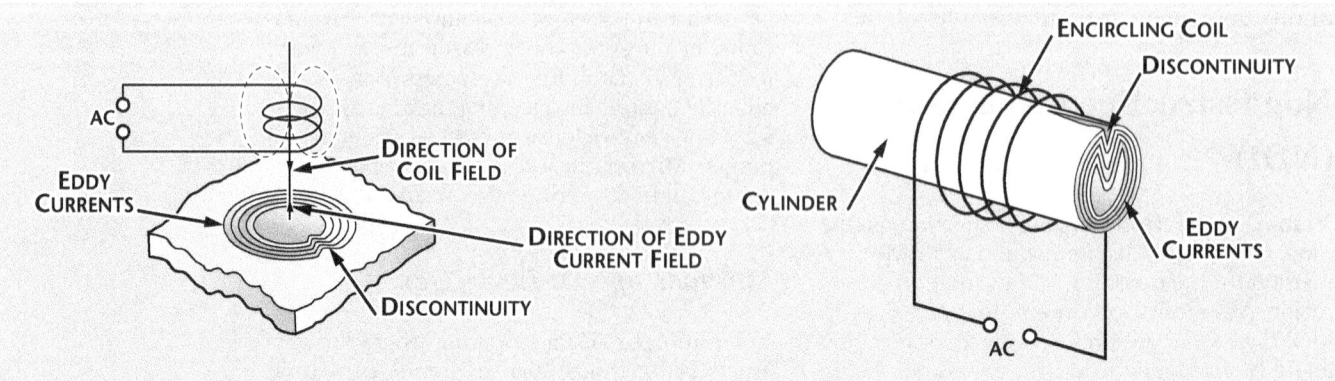
Figure 2-5-5. Eddy current Inspection

required to detect minor or moderate corrosion. Figure 2-5-6 illustrates the X-ray process. Briefly, X-ray works by passing high energy rays generated by an X-ray machine through the material being inspected. This exposes the special film placed on the opposite side of the material. Areas of high density are indicated on the film as underexposed areas, while areas of low density are indicated on the film as overexposed areas.

Proper interpretation of the film will indicate whether defects are present. Moderate to severe corrosion or cracks can be detected using X-ray inspection. This method, like other NDI methods, requires a qualified and certified operator to obtain reliable results.

Ultrasonic inspection. Ultrasonic testing provides a sensitive detection capability for corrosion damage when access is available to a surface with a continuous bulk of material exposed to the corrosion. Figure 2-5-7 shows the principles involved. Ultrasonic inspection is commonly used to detect exfoliation, stress-corrosion cracks, and general material thinning. Ultrasonic digital 2-thickness gauges are not reliable for determining moderate or severe damage prior to removing the corrosion. Highly trained personnel should conduct the examination if any useful information is to be derived from the indicating devices.

Acoustic emission testing. This method using heat-generated emissions can be used to detect corrosion and moisture in adhesive-bonded metal honeycomb structures. Acoustic emission testing can detect corrosion initiation as well as advanced corrosion.

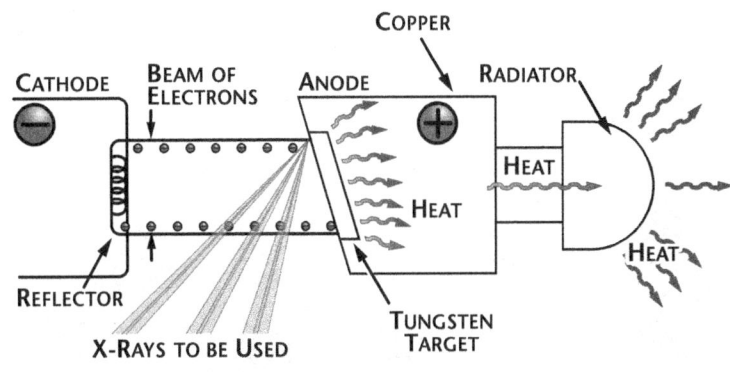

Figure 2-5-6. X-ray inspection

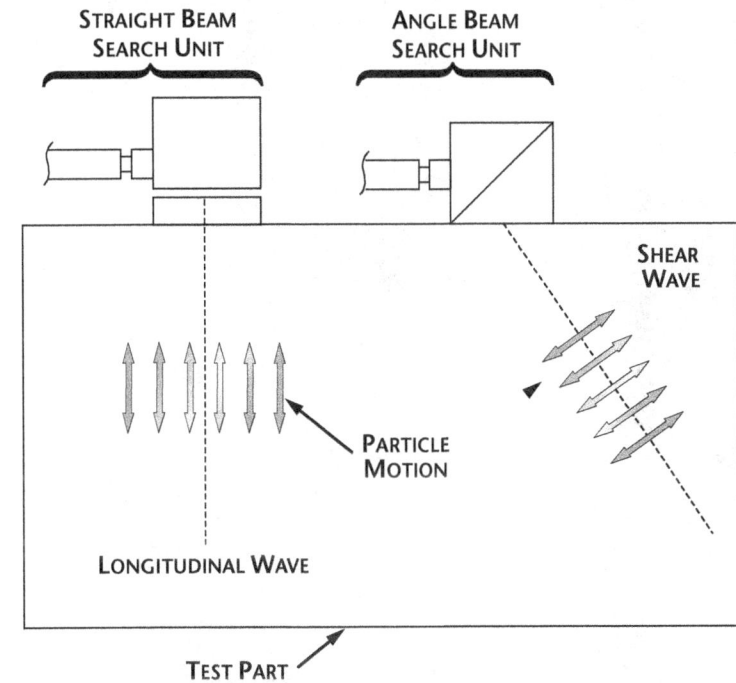

Figure 2-5-7. Ultrasonic inspection

Chapter 3
CORROSION *control*

Corrosion is the electrochemical deterioration of a metal because of its chemical reaction with the surrounding environment. While new and better materials are continuously being developed, this progress is offset by a more aggressive operational environment. The fact that corrosion is a complex phenomenon compounds the problem. It can take many different forms and the resistance of aircraft materials to corrosion can drastically change with only a small environmental change.

Corrosion is a slow process of material deterioration, taking place over a significant period of time. Examples include general corrosion, pitting, and exfoliation. In other forms of corrosion, degradation can occur very quickly, in days or even hours, with catastrophic results. These forms, such as stress corrosion cracking, environmental embrittlement, and corrosion fatigue, depend on both the chemical and mechanical aspects of the environment and can cause catastrophic structural failure without warning.

When active corrosion is found, a positive inspection and rework program is necessary to prevent any further deterioration. In general, any rework would involve the cleaning and stripping of all the finish from the corroded area, removal of corrosion products, and restoration of the surface protective film. Repair of corrosion damage includes removal of all corrosion and corrosion products.

When the corrosion damage is severe and exceeds the damage limits set by the aircraft or parts manufacturer, the part must be replaced. If manufacturer information and limits are not available, then an FAA Designated Engineering Representative (DER) must be consulted before the aircraft or part is returned to service.

Learning Objectives

- Preventive Maintenance
- Types of Cleaning Materials
- Surface Treatment
- Basic Corrosion Removal Techniques
- Corrosion Removal Procedures

Left: Corrosion control begins when the structure is first assembled.

Section 1
Corrosion Control Program

The possibility of an in-flight mishap or excessive down time for structural repairs necessitates an active corrosion prevention and control program. The type and aggressiveness of the corrosion prevention and control program depend on the operational environment of the aircraft. Aircraft exposed to salt air, heavy atmospheric industrial pollution, and/or over water operations will require a more stringent corrosion prevention and control program than an aircraft that is operated in a dry environment.

In order to prevent corrosion, a constant cycle of cleaning, inspection, operational preservation, and lubrication must be followed. Prompt detection and removal of corrosion will limit the extent of damage to aircraft and aircraft components. The basic corrosion prevention and control program should include the following:

- Personnel that are adequately trained in the recognition of corrosion, including conditions, cleaning, treating, and preservation
- A thorough knowledge of corrosion identification techniques
- Proper emphasis on responsibility for corrosion control
- Inspection for corrosion on a scheduled basis
- Washing the aircraft at regularly scheduled intervals
- Routine cleaning or wipe down of all exposed unpainted surfaces
- Keeping drain holes and passages open and functional
- Inspection, removal, and reapplication of preservation compounds on a scheduled basis
- Early detection and repair of damaged protective coatings
- Thorough cleaning, lubrication, and preservation methods at prescribed intervals
- Prompt treatment of corrosion after detection
- Accurate record keeping and reporting of material or design deficiencies
- Use of appropriate materials, equipment, and technical publications

The initial cleaning of an aircraft is shown in Figure 3-1-1.

Preventive Maintenance

"An ounce of prevention is worth a pound of cure" is an understatement where corrosion prevention on aircraft is concerned. Compared to the cost of an aircraft, the cost of corrosion prevention is small. Preventive maintenance is a powerful tool that can be used to effectively control even the most difficult corrosion problem.

Most corrosion prevention programs are adjusted to meet operating conditions. The most common corrosion preventive measures involve keeping the aircraft as clean as possible, all surface finishes intact, correct and timely use of covers and shrouds, periodic lubrication, and the application of preservatives. Years of experience have proven the need for such measures to keep the aircraft airworthy. When corrosion preventive maintenance is neglected, an aircraft soon becomes unsafe to fly.

Surface maintenance. Surface maintenance that includes regular cleaning of the aircraft as well as touch-up of protective paint coatings will prevent corrosion from starting.

It is important that aircraft be kept clean. Cleaning is needed when there is any appreciable amount of soil accumulation within exhaust track areas, or the presence of salt deposits or other contaminants such as stack gases are evident. The presence of excessive oil, exhaust deposits or spilled electrolyte and deposits around battery areas require prompt cleaning.

Figure 3-1-1. Washing the aircraft is the first step in any corrosion control program.

Corrosion Control | 3-3

Figure 3-1-2. The global distribution of corrosion prone areas.

Cleaning is mandatory when any of the following occurs:

- Immediately after exposure to fire-extinguishing materials
- Exposure to adverse weather conditions and salt spray
- Repairs or service which have left stains, smudges, or other gross evidence of maintenance

A daily cleaning or wipe down is required on all exposed unpainted surfaces such as struts, actuating cylinder rods, and so forth.

Aircraft must be thoroughly cleaned before they are stored, and they should be thoroughly cleaned when they are removed from storage.

Components that are critically loaded or designed with minimum safety margins to conserve size and weight, such as helicopter rotor parts and parts that are exposed to corrosive environments, are cleaned as often as possible to minimize exposure to corrosive agents.

> **NOTE:** *Post-cleaning lubrication and preservation of exposed components are necessary to displace any of the cleaning solution entrapped during the cleaning operation.*

Corrosion prevention on the aircraft structure depends on a comprehensive corrosion prevention and control plan that is implemented from the start of operation of an aircraft. That plan should include suggested intervals for cleaning, inspections and repair based on the operating environment. The suggested intervals are based on the geographic location of the aircraft and the type of operation in which it is engaged. The recommended intervals are every 90 days for aircraft in mild corrosion zones. For locations that are considered to be moderate corrosion zones, the interval decreases to every 45 days. In the most severe corrosion zones such as at the seashore and in heavy industrial areas, the inspection interval becomes every 15 days. The maps in Figure 3-1-2 detail the areas of the world where corrosion is most likely to occur.

Prompt corrosion treatment after detection and accurate record-keeping and reporting of material or design deficiencies to the manufacturer and the FAA will contribute to the data that is collected by these parties.

Simple things like keeping drain holes and passages open and functional allow the aircraft to be ventilated and prevent the accumulation of moisture. Sealants, leveling compounds, miscellaneous debris, or corrosion inhibitors should not block drain paths. The replacement of deteriorated or damaged gaskets and sealants to avoid water intrusion and entrapment which leads to corrosion will minimize the exposure of aircraft to adverse environments.

Cleaning Compounds

Cleaning compounds work by dissolving soluble soils, emulsifying oily soils, and suspending solids. There are several types of cleaning compounds, each of which cleans a surface using one or more of these mechanisms.

Highly alkaline cleaning compounds (those with a pH greater than 10) are not recommended. Moderately alkaline cleaners with a pH between 7.5 and 10 are recommended. This type contains detergents, foaming agents, and solvents, and works in the same way as a detergent solution.

High-gloss spot cleaner is recommended for cleaning exhaust track areas of high gloss paint systems. This material contains solvents, detergents, and suspended abrasive matter to remove soil by wearing away the surface that holds it.

Thixotropic (viscous) cleaner is recommended for cleaning wheel wells and replacement of some solvent cleaning where water rinsing can be tolerated. This type of cleaner contains solvents, detergents and some thickening agents. When applied undiluted to an oily or greasy surface, the cleaner clings long enough to emulsify the soil (about 5 to 15 minutes) and is then rinsed away with fresh water.

> **NOTE:** *The use of solvents for cleaning operations is becoming more and more limited due to environmental regulations. Local requirements for waste disposal will determine what types of cleaners may be used.*

Solvent emulsion cleaners work by softening oily soils so that they can be emulsified by the detergent and rinsed away.

Detergent solution cleaners dissolve in water and clean by dissolving soluble salts, emulsifying low viscosity oils, and suspending easily removed dirt and dust. They are not very effective on grease, but are excellent cleaners for interior lightly soiled areas, plastics, and instrument glass covers.

Cleaning solvents dissolve oily and greasy soils so that they can be easily wiped away or absorbed on a cloth. Solvents differ significantly in cleaning ability, toxicity, evaporation rate, effect on paint, and flammability. The use of cleaning solvent is intended for localized spot application only. A dry cleaning solvent conforming to PD-680, Type II, is the most common cleaning solvent used on aircraft, due to its low toxicity, minimal effect on paint, and relative safety. Other solvents, such as alcohols, ketones, chlorinated solvents, and naphtha are specialized materials and have restricted use. Refer to the manufacturer's maintenance and cleaning procedures for specific applications.

> **NOTE:** *More cleaning compound is not necessarily better. Mixing more cleaning compound increases the pH of the solution which can do more harm than good. A cleaning compound should always be mixed in accordance with the manufacturer's recommendations.*

Miscellaneous cleaning agents include:

- Plastic polish containing mild abrasive matter to polish out scratches in canopy materials
- Alkaline chemicals used to neutralize specific acidic soils
- Sodium bicarbonate for electrolyte spills from sulfuric acid batteries
- Monobasic sodium phosphate and boric acid for electrolyte spills from nickel-cadmium batteries

Steam cleaning is not recommended for general use on aircraft. It erodes paint, crazes plastic, debonds adhesives, damages electrical insulation, and drives lubrication out of bearings.

Section 2
Hazardous Materials

A hazardous material is defined as material presenting hazards to personnel, property, or the environment in the course of handling, storing, and using such materials. While hazardous materials can be used safely, extra precautions are needed when handling and storing these materials.

> **NOTE:** *Hazardous materials, such as chemicals, require a hazardous chemical or material identification label, shown in Figure 3-2-1.*

The standard symbol format uses numerals and symbols to describe the degree of hazard with respect to health, fire, reactivity, and the specific hazard of the packaged product.

Flammable and Combustible Liquids

Combustible liquids are defined as liquids having flash points at or above 100°F (37.8°C). Flammable liquids are defined as liquids having flash points below 100°F (37.8°C). Fire is a very serious hazard, but breathing toxic fumes in unventilated spaces creates an equally hazardous circumstance.

Flash point is the minimum temperature at which a liquid gives off a vapor within a test vessel, in sufficient concentration to form an ignitable mixture with air near the surface of the liquid.

Solvents

Solvents are liquids that dissolve other substances. They are used in many products such as paints and degreasing fluids. There are many types of solvents, but for aircraft cleaning purposes, organic solvents, such as aircraft cleaning compounds are most often used. Aside from posing a fire hazard, inhaling the vapors can seriously affect the brain and the central nervous system. Therefore, solvents are to be used only in well-ventilated spaces. Gloves as well as a face shield should be worn to protect skin and eyes. An approved respirator must be used to prevent breathing toxic vapors. If gloves are not worn, the skin will lose its fatty protection and will dry, crack, and become infected.

NFPA
- This label is for emergency response and fire fighters.
- Hazard rating is from 0 (no hazard) to 4 (extreme).

FLAMMABILITY (FLASH POINTS)
0 = Will not burn
1 = Above 200°F
2 = Between 100 - 200°F
3 = Below 100°F
4 = Flash Point Below 73°F

SPECIFIC HAZARD
ACID - acid
ALK - alkali
COR - corrosive
OX - oxidizer
P - polymerization
☢ - radioactive
W̶ - use no water

HEALTH HAZARD
0 = Normal Material
1 = Slight Hazard
2 = Moderately Hazardous
3 = Extremely Hazardous
4 = Deadly

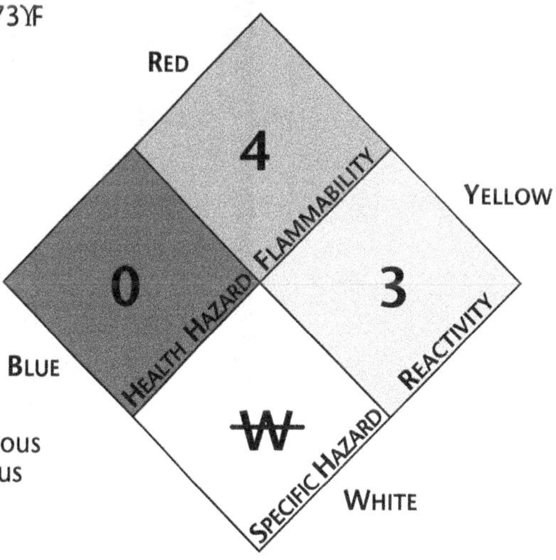

REACTIVITY
0 = Stable
1 = Unstable if heated
2 = Violent Chemical Change
3 = Shock or heat may detonate
4 = Rapidly capable of detonation or Explosion

Figure 3-2-1. Knowledge of the various parts of the hazardous material (HAZMAT) symbol is important.

When solvents contain more than 24% chlorinated materials by volume, they must be kept in specially marked containers. The equipment in which the solvent is to be used should be designed and operated to prevent escape of the solvent. All personnel who work near chlorinated solvents should be careful to avoid breathing the vapors. While the vapors from some solvents are more toxic than others, prolonged breathing of any fumes can be very hazardous to health.

Containers holding paints, lacquers, removers, thinners, cleaners, or any volatile or flammable liquids must be kept tightly closed when not in use. All flammable and volatile liquids must be stored in a separate building or a flammable liquids storage cabinet similar to the one in Figure 3-2-2.

The storage area must be well-ventilated and located where its contents will not be exposed to excessive heat, sparks, flame, or direct rays of the sun. Enclosed storage areas must also have a fixed CO_2 or Halon extinguishing system. All electrical fixtures, outlets, and other wiring must be of the explosion-proof class. Wiping rags and other flammable waste material must always be placed in tightly closed containers that are emptied at the end of the work shift.

When flammable and volatile materials are stored in a paint room, the temperature inside could become very high, especially during the summer months. As the temperature increases, liquids expand. In the past, maintenance personnel have received serious chemical burns on the face, hands, and arms due to opening a hot can of solvent. This hazard increases many times when working with the more volatile types of liquids, such as paint strippers. Before opening a container of solvent that has been stored in a high temperature area, it should be cooled down. Use common sense around flammable and volatile liquids.

Another hazard associated with solvents is their effect on the surface of the material being cleaned. Some solvents, such as methyl ethyl ketone (MEK) and toluene, will damage rubber, synthetic rubber, asphalt coverings, etc. This damaging effect must be considered when selecting cleaning materials. Cleaning materials may do a good job in removing dirt, grease, oil, and exhaust gas deposits, but they may also damage the object being cleaned or soften and ruin an otherwise good paint coating.

Types of Cleaning Materials

Dry cleaning solvent. This material is a petroleum distillate commonly used in aircraft cleaning. It is used as a general all-purpose cleaner for metals, painted surfaces, and fabrics. It may be applied by spraying, brushing, dipping, and wiping.

Aliphatic naphtha. This is an aliphatic hydrocarbon product used as an alternate compound for cleaning acrylics and for general cleaning purposes when fast evaporation and no film residue is required. It may be applied by dipping and wiping. Saturated surfaces must not be rubbed vigorously. Aliphatic naphtha must not be used with a synthetic wiping cloth that will build up a static charge and create a spark. It is a highly volatile and flammable solvent that has a flash point below 80°F, and should be used only in well ventilated areas.

Safety solvent. Safety solvent is used for general cleaning and grease removal from assembled and disassembled engine components in addition to spot cleaning. It should not be used on painted surfaces. Safety solvent is not suitable for oxygen systems, although it may be used for other cleaning in ultrasonic cleaning devices. It may be applied by wiping, scrubbing, or booth spraying. The term safety solvent is derived from its high flash point.

Figure 3-2-2. All flammables should be stored properly

Methyl ethyl ketone (MEK). MEK is used as a cleaner for bare metal surfaces and areas where sealants are removed. Normally, MEK is applied over small areas using wiping cloths or soft bristle brushes.

> **CAUTION:** *Avoid prolonged breathing and skin contact. MEK should be used only in well ventilated spaces. Extreme care should be used when working around transparent plastics as MEK will damage them upon contact.*

Sodium bicarbonate. Sodium bicarbonate is used as a neutralizing agent on sulfuric acid battery electrolyte deposits.

Sodium bicarbonate may also be used to neutralize urine deposits. It is applied with a sponge using a mixture of 8 ounces of sodium bicarbonate to 1 gallon of fresh water. The area is then flushed with fresh water.

Sodium phosphate. Sodium phosphate is used to neutralize electrolyte spills from nickel cadmium batteries.

Aircraft Surface Cleaning Compound

Water emulsion cleaners are used to clean aircraft. These cleaners tend to disperse contaminants into tiny droplets that are held in suspension in the cleaner until they are flushed from the surface. Water emulsion compounds contain emulsifying agents, coupling agents, detergents, solvents, corrosion inhibitors, and water. These compounds are intended for use on painted and unpainted surfaces in heavy duty cleaning operations, when milder specification materials of lower detergency would not be effective. They are used in varying concentrations, depending upon the condition of the surface.

Water emulsion cleaner is applied by starting at the bottom of the area being cleaned. The mixed solution is applied by spraying or brushing to avoid streaking. Surface soils are loosened by mild brushing or mopping. Then, the surface is given a thorough freshwater rinse.

Alkaline water base cleaner compound. Alkaline water base cleaners are similar to the water emulsion cleaner. They are a general purpose cleaner used to remove light and moderate soil. Alkaline water base cleaner compound is safe for use on fabrics, leather, glass, ceramics, and transparent plastics. When this cleaner is used, the procedure described by the manufacturer should be used.

Liquid detergents. Detergents are used to clean transparent and acrylic plastics and cockpit indicator glass covers. They are also used at intermediate-level maintenance activities as a water-based solvent spray in cleaning booths and aqueous ultrasonic cleaners for removing contaminants.

Isopropyl alcohol. Isopropyl alcohol is a general-purpose cleaner and solvent. It is used to remove salt residue and contaminants from avionics and electrical equipment. When it is used as a cleaner on electrical contact surfaces, it should be applied as a solution of 1 part de-ionized or distilled water to 1 part isopropyl alcohol. An acid brush or pipe cleaner should be used to apply it. After application it should be wiped clean and air dried.

> **NOTE:** *Isopropyl alcohol is highly flammable and requires the same handling and storage procedures as other solvents.*

Section 3
Cleaning Procedures

The following general cleaning procedures are recommended:

1. Remove or disconnect all electrical power and ground the aircraft
2. Protect against water or cleaning compound intrusion by closing doors, openings, cover vents, pitot static openings, and covering wheels
3. Accomplish pre-wash lubrication. Lubricate in accordance with applicable maintenance compounds
4. Mix cleaning solution to manufacturer's manual and recommendations
5. Use spray not a stream of water during aircraft wash
6. Do not use abrasive cleaning pads
7. Rinse aircraft with fresh water to remove all cleaning compounds

The following post-cleaning procedures are recommended:

1. Remove all covers, plugs and masking materials
2. Inspect and clear all drain holes
3. Inspect open and all known water trap areas for water accumulation and proper drainage
4. Lubricate aircraft in accordance with applicable maintenance manual
5. Apply operational preservation

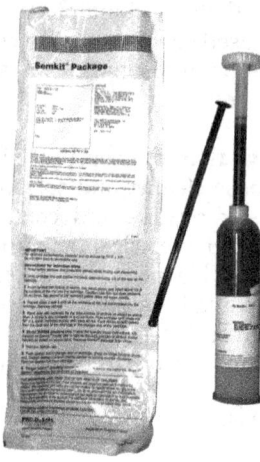

Figure 3-3-1. Two types of commonly used sealants

Preservation

The day-to-day application of corrosion preventive compounds is used to protect metal aircraft parts and components. They function by preventing corrosive materials from contacting and corroding bare metal surfaces. Many of these compounds are also able to displace water and other contaminants from the surfaces to be protected. Some also provide lubrication as well as corrosion protection. Corrosion preventive compounds vary in appearance and consistency from thick, black types to light oils. Thicker compounds provide the best protection and are more difficult to remove. They do not crack, chip, or peel but provide continuing protection applied by brush, sponge or stick. The lighter types are a yellow to gold color while others are brown. The lighter oils may have a water displacing quality and provide a thin film of lubrication. All types of coatings must be removed and replaced regularly to provide continued protection.

Surface Treatment

An important step in the corrosion control process is the surface treatment of the metal with a prescribed chemical to form a protective film. Chemical surface treatments that are properly applied provide corrosion resistance to the metal and improve the adhesion of subsequently applied paints. These surface treatments, also known as chemical conversion coatings, chromate conversion coatings, chemical films, or pre-treatments, are aqueous acid solutions of active inorganic compounds which convert aluminum or magnesium surfaces to a corrosion resistant film.

The surface should be prepared for application of the chemical conversion coatings by feathering the edges of the areas that have been chemically stripped prior to pretreatment and repainting to ensure a smooth, overlapping transition between the old and new paint surfaces.

- Clean the area with a fine or very fine abrasive mat saturated with water.

- Rinse by flushing with fresh water. Particular attention should be given to fasteners and other areas where residues may become entrapped. At this stage in the cleaning, the surface should be water break-free. A surface showing water breaks (water beading or incomplete wetting) is usually contaminated with grease or oil which will later interfere with conversion coating, sealing, and painting.

- If the surface is not free of water breaks, clean the area again using a new abrasive mat and rinse thoroughly with water.

Sealants

Sealants are one of the most important tools for corrosion prevention and control. They prevent the intrusion of moisture, salt, dust, and aircraft fluids, which can lead to extensive corrosion. For sealants to be effective, it is critical that the correct sealant be chosen for a specific area or situation and that it be applied correctly. Figure 3-3-1 shows some of the different types of sealants that can be used in aircraft.

Sealants are used in the following applications:

- Fuel tank sealing
- Pressure areas
- Weather sealing
- Firewalls
- Electrical
- Acid-resistant areas
- Windows
- High temperature applications
- Aerodynamic sealing

There are numerous sealing compounds available with different properties and intended use. Refer to the aircraft manufacturer's manual for specific information concerning selection of the sealing compound and proper application. Observe the warning and cautions of the manufacturer when using sealing compounds. Sealing compounds generally are divided into two major types, those requiring a curing agent and those which cure in air.

Polysulfide, polythioether, and polyurethane sealing compounds consist of the prepolymer base and the accelerator or curing agent. When thoroughly mixed, the catalyst cures the base to a rubbery solid. Rates of cure depend on the type of base, catalyst, temperature, and humidity. A full cure may not be achieved for as long as seven days.

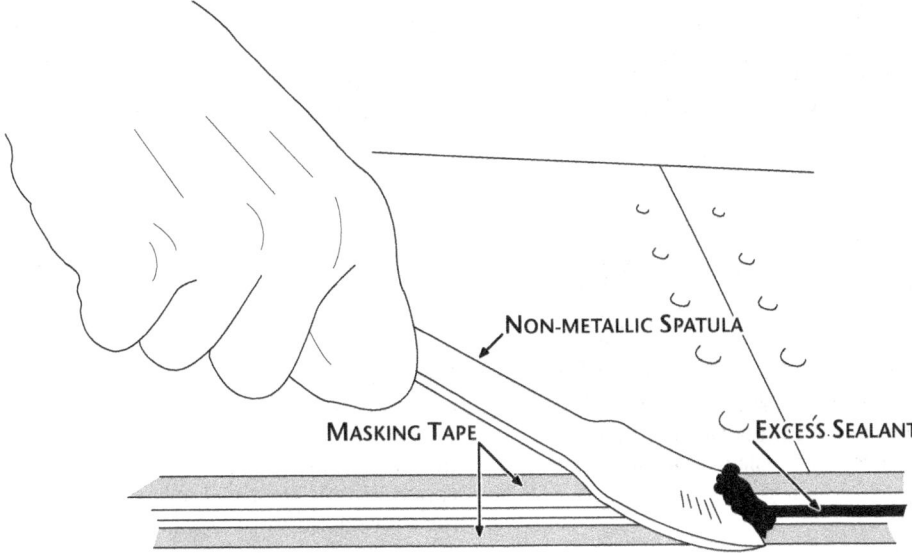

Figure 3-3-2. A non-metallic scraper is the best tool for spreading sealants.

Silicone sealing compounds generally consist of one component which cures by reaction with moisture in the air. If silicones are applied too thick or in such a way as to prevent moisture from entering the material, they may not cure at all.

NOTE: *Many silicone sealing compounds produce acetic acid (vinegar smell) while curing, which can lead to severe corrosion problems. The use of silicone sealing compounds on aircraft should be limited to those non-corrosive products.*

Some sealing compounds may require the application of a special primer or adhesion promoter before sealant application in order to develop a good adhesive bond with the surface. Use only those primers or adhesion promoters recommended by the product manufacturer.

Following the removal of corrosion and application of a chemical surface treatment all surfaces, except fuel tank interiors, should be primed. If the surfaces have been contaminated following surface treatment, clean the area with a cleaning solvent and a clean cloth. Dry the surfaces immediately with a clean cloth and do not allow solvent to evaporate from the surface.

Mask off the area being sealed to prevent sealant from contacting adjacent areas during application and post-application smooth-out operations. Examples of where masking may be beneficial are fillet sealing of exterior surface lap and butt seams.

When required by the manufacturer, apply a thin coating of an adhesion promoting solution. Allow to dry by evaporation without touching the area for 30 minutes to 1 hour before applying sealant.

Spatula type sealants may be applied with a non-metallic spatula or scraper. Avoid the entrapment of air. Work sealant into recesses by sliding the edge of the scraper firmly back over the recesses between the tape. Smoothing will be easier if the non-metallic scraper is dipped in water, as shown in Figure 3-3-2.

Sealant that is to be applied with a brush is applied and smoothed until the desired thickness is reached. Sealants that are applied with a caulking gun will not usually require masking and are especially adaptable to filling seams or the application of form-in-place gaskets.

Sealant that can be applied with a spray gun should be applied in a solid, continuous pattern.

Allow sealant to dry or cure to manufacturer's recommendations.

When required, prime sealant as soon as it no longer feels tacky, then topcoat as necessary. This is shown in Figure 3-3-3.

Figure 3-3-3. The primer and topcoat should be sprayed on as soon as possible.

Photo courtesy of JRA Executive Air

Figure 3-3-4. The reinforcement patches around these two fitting attachments were injection-sealed with a Semkit® type sealant. Fillet sealing is a necessary part of corrosion control.

Faying surfaces. Faying surface sealant is applied between the contacting surfaces of two or more parts, and is the most effective seal that can be produced. It should be used for all assembly and, where possible, reassembly. There are two types of faying surface seal installations, removable and permanent. Figure 3-3-4 is an example of a properly sealed faying surface.

- The removable type is for access doors, removable panels, inspection plates, windows, etc. The sealant is applied to the substructure and a parting agent applied on the removable panel during cure.
- The permanent type is for sealing between parts of a structure that are permanently fastened together.

Fillet or seam seals. The fillet, or seam seal is the most common type found on an aircraft. Fillet seals are used to cover structural joints or seams along stiffeners, skin butts, walls, spars, and longerons, and to seal around fittings and fasteners. It should be used in conjunction with faying surface sealing and in place of it if the assembly sequence restricts the use of faying surface sealing.

Injection sealing. Injection sealing primarily fills voids created by structural joggles, gaps, and openings. Only those sealants recommended by the aircraft/equipment manufacturer should be used.

The sealant is forced into the area using a hand or pneumatic sealant gun. This produces a continuous seal where it becomes impossible to lay down a continuous bead of sealant while fillet sealing. Clean the voids of all dirt, chips, burrs, grease, and oil before injection sealing. A vent hole may be needed to allow air to escape from a void as the sealant is injected. Failure to do so can force the surfaces apart, or only partially fill the area.

Fastener sealing. The fastener sealing method depends on the type of fastener. Fasteners are sealed either during or after assembly. To seal a permanent fastener during assembly, apply the sealant to the hole or dip the fastener into the sealant, and install the fastener while sealant is wet. For removable fasteners, start the fastener in the hole and apply sealant to the lower side of the fastener head or countersink. To seal after assembly, apply sealant to the fastener head after installation.

Fuel tanks. Sealing of fuel cells should be accomplished in accordance with the aircraft manufacturer's maintenance manual procedures. Some sealants are not fuel resistant and can contaminate the fuel system when they deteriorate. Improper sealing of joints in a fuel tank results in fuel leaks that create a potentially hazardous situation.

Paint Finishes and Touch-up Procedures

The primary objective of any paint system is to protect exposed surfaces against corrosion and other forms of deterioration. Uses for particular paint schemes include:

- Identification markings
- Abrasion protection
- Specialty coatings (i.e., walkway coatings).
- High visibility requirements

The paint system on aircraft consists of a primer coat and a topcoat. The primer promotes adhesion and contains corrosion inhibitors. The topcoat provides durability to the paint system, including weather and chemical resistance, along with the coloring necessary for operational requirements.

Some aircraft surfaces (Teflon-filled, rain erosion, walkways, etc.) require specialized coatings to satisfy service exposure and operational needs. For these surfaces, refer to the specific manufacturer's maintenance manual for the aircraft in question.

NOTE: *The Environmental Protection Agency, as well as certain local air pollution control districts, have implemented rules which limit the volatile organic content (VOC), or solvent content, of paints applied to aircraft. It is the responsibility of the user to insure that these rules are understood and obeyed. Failure to comply with current rules can result in large fines.*

Much of the effectiveness of a paint finish and its adherence depend on careful preparation of the surface before touch-up and repair.

Aged paint surfaces must be scuff sanded to ensure adhesion of over-coated, freshly applied paint. Sanding requires a complete roughening of the paint surface and can be accomplished by hand sanding or with the use of power tools.

For final preparation, ensure that surfaces to be painted are free of corrosion, have been prepared and the surrounding paint feathered, and have been conversion coated. Replace any seam sealants when necessary. Mask areas as required to prevent overspray.

Touch-up, over-coat and total repaint. Primers should be thinned with the applicable thinner as recommended by the paint manufacturer, stirred, and applied in even coats. Primer thickness varies for each type primer but generally the total dry film thickness is 0.6 to 0.9 mils (0.0006 to 0.0009 inch). You should be able to see through this film thickness. Allow primer to air dry before topcoat application in accordance with the paint manufacturer's recommendations. Normally topcoat application should occur within 24 hours after primer application. The painting sequence is important for complete coverage of the aircraft, the method in Figure 3-3-5 is normally used.

Topcoats should be thinned with the applicable thinner as recommended by the paint manufacturer, stirred, and applied in even coats. Topcoat thickness varies for each topcoat but generally; the total dry film thickness is 1.5 to 2.0 mils (0.0015 to 0.002 inch). Allow the topcoat to air dry in accordance with the paint manufacturer's instructions.

Teflon-filled or anti-chafe coatings should be applied over a primer in accordance with the manufacturer's instructions.

Walkway compounds should be applied over a primer in accordance with the manufacturer's instructions.

Section 4
Basic Corrosion Removal Techniques

All corrosion products should be removed completely when corroded structures are reworked. Before starting rework of corroded areas, carry out the following:

1. Document corrosion damage
2. Position the aircraft in a wash rack or provide washing apparatus for rapid rinsing of all surfaces

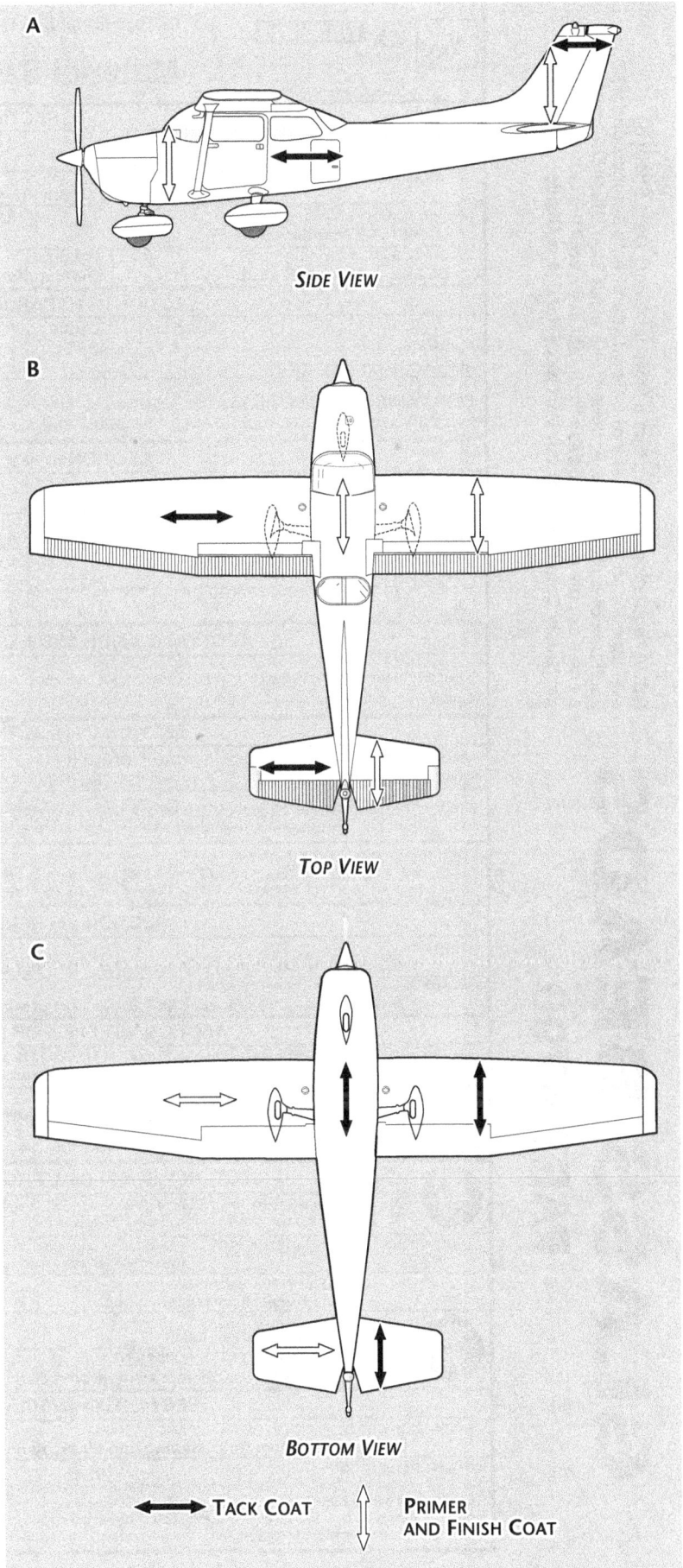

Figure 3-3-5. A painting sequence for a light single-engined airplane

OSHA regulations require that a Material Safety Data Sheet be available for all users. Material Safety Data Sheets are included with each box of our abrasive. Included with most Skat Blast Cabinets is a small supply of Glass Beads for initial start-up. This is the Material Safety Data Sheet for our Part No. 6700 Glass Beads.

GLASS BEADS PART 6700

WARNING — Glass Beads, if spilled on the floor, are as slippery as ice. Sweep up spills promptly.

Occupational Safety and Health Administration
Material Safety Data Sheet

PART NO. 6700
GLASS BEADS
Revised 9-14-98

SECTION I

DISTRIBUTOR NAME: Distributor---Skat Blast, Inc.
EMERGENCY PHONE NO. (330) 533-9477
ADDRESS, (Number, Street, City, State and Zip Code): 7077 State Rt. 446, Canfield, OH 44406
CHEMICAL NAME AND SYNONYMS: Glass (CAS Number 65997-17-3)
TRADE NAME AND SYNONYMS: GLASS BEADS
CHEMICAL FAMILY: GLASS, SODA-LIME TYPE
FORMULA: Sodium Aluminum Silicate base Glass

SECTION II - HAZARDOUS INGREDIENTS

	OSHA-PEL	ACGIH-TLV	ALLOYS AND METALLIC COATINGS
NUISANCE DUST	15mg/m^3	10mg/m^3	
NUISANCE DUST - RESPIRABLE	5mg/m^3	5mg/m^3	N/A

CONTAINS NO FREE SILICA - All components are amorphous and non-crystalline; none are SARA Title III reportable.
NUISANCE DUST - to the best of our knowledge, this material is non-hazardous as per OSHA 29 CFR 1910.1200.

SECTION III - PHYSICAL DATA

BOILING POINT (C)	MELTING POINT (C)	SPECIFIC GRAVITY (@ 60 degrees F)	
N/A	Above 1100°F		2.4-2.6g/cm^3
VAPOR PRESSURE (mm Hg)	N/A	PERCENT SOLID BY WEIGHT	N/A
VAPOR DENSITY (Air = 1)	N/A	EVAPORATION RATE	N/A
SOLUBILITY IN WATER	None	APPEARANCE AND COLOR	White, odorless, tasteless powder.

SECTION IV - FIRE AND EXPLOSION HAZARD DATA

Flash Point (Method Used): None. Flammable Limits: None. Extinguishing Media: Does not burn. Water - avoid creating dust.
Materials to avoid - (Incapability) Concentrated Hydrofluoric Acid, fluosilicic, phosphoric acids, and hot, strong alkaline solutions.

SECTION V - HEALTH HAZARD DATA

ROUTES OF ENTRY: Inhalation, Ingestion
CARCINOGENICITY - This product is not listed as a potential carcinogen in either the NIP, IARC, or OSHA.

HEALTH HAZARD - Repeated or prolonged inhalation of dust in excess of permissible exposure limits may result in irritation to the respiratory tract.
OVEREXPOSURE - Can aggravate existing respiratory conditions and eye irritation.

EMERGENCY AND FIRST AID PROCEDURE:
Eye Contact: Flush with running water for 15 min. If irritation or redness persists, see a physician. Skin Contact: Wash area well with soap and water. Ingestion: Seek medical help if large quantities of material have been ingested.

SECTION VI - REACTIVITY DATA

STABILITY		CONDITIONS TO AVOID
Stable.	Hazardous Polymerization - Will not occur	Excessive dust, slippery if spilled

HAZARDOUS DECOMPOSITION PRODUCTS
None — Glass Beads will break down into progressively smaller particles during normal use.

SECTION VII - SPILL OR LEAK PROCEDURES

STEPS TO BE TAKEN IN CASE MATERIAL IS RELEASED OR SPILLED. Considered as non-hazardous per EPA 29CFR1910.1200. Sweep from floor to prevent slipping hazard. Wear NIOSH-approved respirator. If respiratory aggravation, go to a well-ventilated area.

WASTE DISPOSAL METHOD
Collected dust from blast cleaning or shot peening operations always contain contaminates from the surface of the parts being processed; and, therefore, the dust may be classed as a hazardous waste and, as such, must be disposed of according to appropriate Local, State, or Federal regulations. The RCRA status of UNUSED material is non-hazardous, according to the list of CERCLA chemicals.

SECTION VIII - SPECIAL PROTECTION INFORMATION

RESPIRATORY PROTECTION (SPECIFY TYPE) Use NIOSH/OSHA-approved respiratory equipment. Positive pressure air-supplied type recommended. If beads or dust cause eye irritation, flush eye(s) with water or eye wash.

VENTILATION: N/A
LOCAL EXHAUST: Follow OSHA standards; use adequate dust collecting system to remove suspended particulate from equipment and ambient environment.
MECHANICAL (GENERAL): As required for nuisance dust.
OTHER: Provide eyewash station in the area.
PROTECTIVE GLOVES: As required per job.
EYE PROTECTION: Normal for dust---use safety goggles.
OTHER PROTECTIVE EQUIPMENT: If blasting, use appropriate protective clothing, air-supplied hood/respirator & other safety equipment.

SECTION IX - SPECIAL PRECAUTIONS

PRECAUTIONS TO BE TAKEN IN HANDLING AND STORING
Observe maximum floor loading and stacking limitations due to density of product.

OTHER PRECAUTIONS
Keep dry. Store material away from incompatible materials. Avoid generating dust.
The company has no control over this product or its use after it leaves our facility. The company assumes no liability for loss or damage from the proper or improper use of this product. The information presented here has been compiled from sources considered to be reliable and accurate to the best of our knowledge and belief, but is not guaranteed to be so. Revision Date, September 14, 1998.

Rev 3/28/00

Figure 3-4-1. This MSDS represents the information for the basic product

3. Connect a static ground line from the aircraft to a grounding point
4. Prepare the aircraft for safe ground maintenance
5. Install all applicable safety pins, flags, and jury struts
6. Protect the pitot-static ports, louvers, airscoops, engine opening, wheels, tires, magnesium skin panels, and airplane interior from moisture and chemical brightening agents
7. Protect the surfaces adjacent to rework areas from chemical paint strippers, corrosion removal agents, and surface treatment materials

Safety Precautions

General safety precautions for handling materials with hazardous physical properties must be followed. Those precautions must also address emergency procedures for immediate treatment of personnel who have inadvertently come into contact with harmful materials. All personnel responsible for using or handling hazardous materials should be thoroughly familiar with the information in the Material Safety Data Sheets (MSDS) for the products. A MSDS is shown in Figure 3-4-1.

Chemical precautions. When required to use or handle solvents, special cleaners, alkalis or acid paint strippers, etchants which are corrosion removers containing acids, or surface activation materials like Alodine 1200, the following safety precautions should be observed:

- Avoid prolonged breathing of solvent or acid vapors
- Never add water to acid. Always add acid to water
- Mix all chemicals per the manufacturer's instructions
- Avoid prolonged or repeated contact of solvents with the skin, cleaners, etching or conversion coating material. Goggles or plastic face shields and suitable protective clothing should be worn when cleaning, stripping, etching, or conversion coating of aircraft surfaces.
- When mixing alkalis with water or other substance, use containers that are made to withstand the heat generated by this process
- Wash any paint stripper, etching, or conversion coating material immediately from body, skin, or clothing
- Equipment should be effectively grounded where any flammable materials are being used
- Do not use solvents with a low flash point, (below 100°F) such as methyl ethyl ketone (MEK) and acetone in enclosed areas
- Check and follow all applicable restrictions and requirements on the use of solvents, primers, and top coats and disposal of waste material

Blasting precautions. The following precautions should be taken when using any type of blasting equipment:

- Operators should be adequately protected with complete face and head covering equipment, and provided with pure breathing air
- Static-ground the dry abrasive blaster and the material being blasted
- Magnesium cuttings and small shavings can ignite easily and are an extreme hazard. Fires of this metal must be extinguished with a Class D for metals, fire extinguisher
- Titanium alloys and high-tensile strength steel create sparks during dry abrasive blasting. Care should be taken to ensure that hazardous concentrations of flammable vapors do not exist.

Corrosion Removal Procedures

The effectiveness of corrosion control depends on how well basic work procedures are followed. The following are common work practices:

- If rework procedures or materials are unknown, contact the aircraft manufacturer or FAA authorized Designated Engineering Representative (DER) before proceeding.
- The work areas, equipment, and components should be clean and free of chips, grit, dirt, and foreign materials as shown in Figure 3-4-2.
- Do not mark on any metal surface with a graphite pencil or any type of sharp, pointed instrument. Temporary markings (defined as markings soluble in water) should be used for metal layout work or marking on the aircraft to indicate corroded areas.
- Graphite should not be used as a lubricant for any component. Graphite is cathodic to all structural metals and will generate galvanic corrosion in the presence of moisture, especially if the graphite is applied in dry form.
- Do not abrade or scratch any surface unless it is an authorized procedure. If surfaces

Figure 3-4-2. Removal of oil and surface dirt with bristle brush

are accidentally scratched, the damage should be assessed and action taken to remove the scratch and treat the area.

- Coated metal surfaces should not be polished for aesthetic purposes. Buffing would remove the protective coating and a brightly polished surface is normally not as corrosion resistant as a non-polished surface unless it is protected by wax or paint.

- Protect surrounding areas when welding, grinding, or drilling, to prevent contamination with residue from these operations. In those areas where protective covering cannot be used, remove the residue by cleaning.

- Severely corroded screws, bolts, and washers should be replaced.

General guidelines. All depressions resulting from corrosion rework should be faired or blended with the surrounding surface.

Remove rough edges and all corrosion from the damaged area. All dish-outs should be elliptically shaped with the major axis running spanwise on wings and horizontal stabilizers, longitudinally on fuselages, and vertically on vertical stabilizers.

In critical and highly stressed areas, all pits remaining after the removal of corrosion products should be blended out to prevent stress risers that may cause stress corrosion cracking. On a non-critical structure, it is not necessary to blend out pits remaining after removal of corrosion products by abrasive blasting, since this results in unnecessary metal removal.

Rework small pits and depressions by forming smoothly blended dish-outs, using a ratio of 20:1, length to depth as is shown in Figures 3-4-3 and 3-4-4.

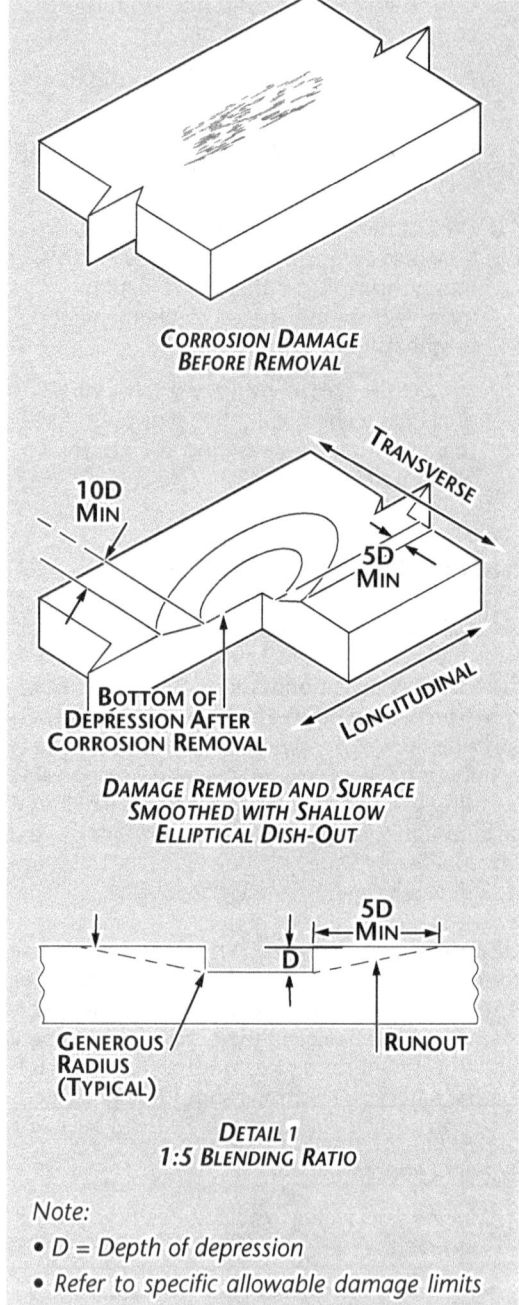

Note:
- D = Depth of depression
- Refer to specific allowable damage limits for maximum allowable depth.
- Since maximum depth varies at different locations, maximum size of dish-out will also vary.
- The blending ratio shall be maintained at all times unless otherwise specified in a specific repair.
- See detail 1 for example of blending.

Figure 3-4-3. Blend-out of pits in corroded area

In areas having closely spaced multiple pits, intervening material should be removed to minimize surface irregularity or waviness.

Steel nut-plates and steel fasteners should be removed before blending corrosion out of an aluminum structure. Steel or copper particles embedded in aluminum can become points of future corrosion. All corrosion products must be removed during blending to prevent reoccurrence of corrosion.

Removal by blasting. Abrasive blasting is a process for cleaning or finishing ferrous metals by directing a stream of abrasive particles against the surface of the parts. Abrasive blasting is used for the removal of rust and corrosion and for cleaning before painting or plating. The part to be blast-cleaned should be removed from the aircraft, when possible. When removal is not possible, areas adjacent to the part should be masked or protected from abrasive impingement and contamination.

- Parts should be dry and clean of oil, grease or dirt prior to blast cleaning.
- Close-tolerance surfaces, such as bushings and bearing shafts, should be masked.
- Blast-clean only enough to remove corrosion coating. Proceed immediately with surface treatments as required.

Chemical cleaners. Use chemicals with great care when cleaning assembled aircraft. The danger of entrapping corrosive materials in faying surfaces and crevices counteracts any advantages in their speed and effectiveness.

Use materials that are relatively neutral and easy to remove.

Removal of spilled battery acid. The battery, battery cover, battery box and adjacent areas will corrode if battery acid spills onto them. To clean spilled battery acid, brush off any salt residue and sponge the area with fresh water. For lead-acid batteries, sponge the area with a solution of six ounces of sodium bicarbonate (baking soda) per gallon of fresh water. Apply generously until bubbling stops and let the solution stay on the area for five to six minutes, but do not allow it to dry.

For nickel-cadmium batteries, sponge the area with a solution of six ounces of monobasic sodium phosphate per gallon of fresh water. Sponge area again with clean fresh water and dry surface with compressed air or clean wiping cloths.

Standard Methods of Corrosion Removal

Several standard mechanical and chemical methods are available for corrosion removal. Mechanical methods include hand sanding using abrasive mat, abrasive paper, or metal wool. Powered mechanical methods such as sanding, grinding, buffing, abrasive mat, grinding wheels,

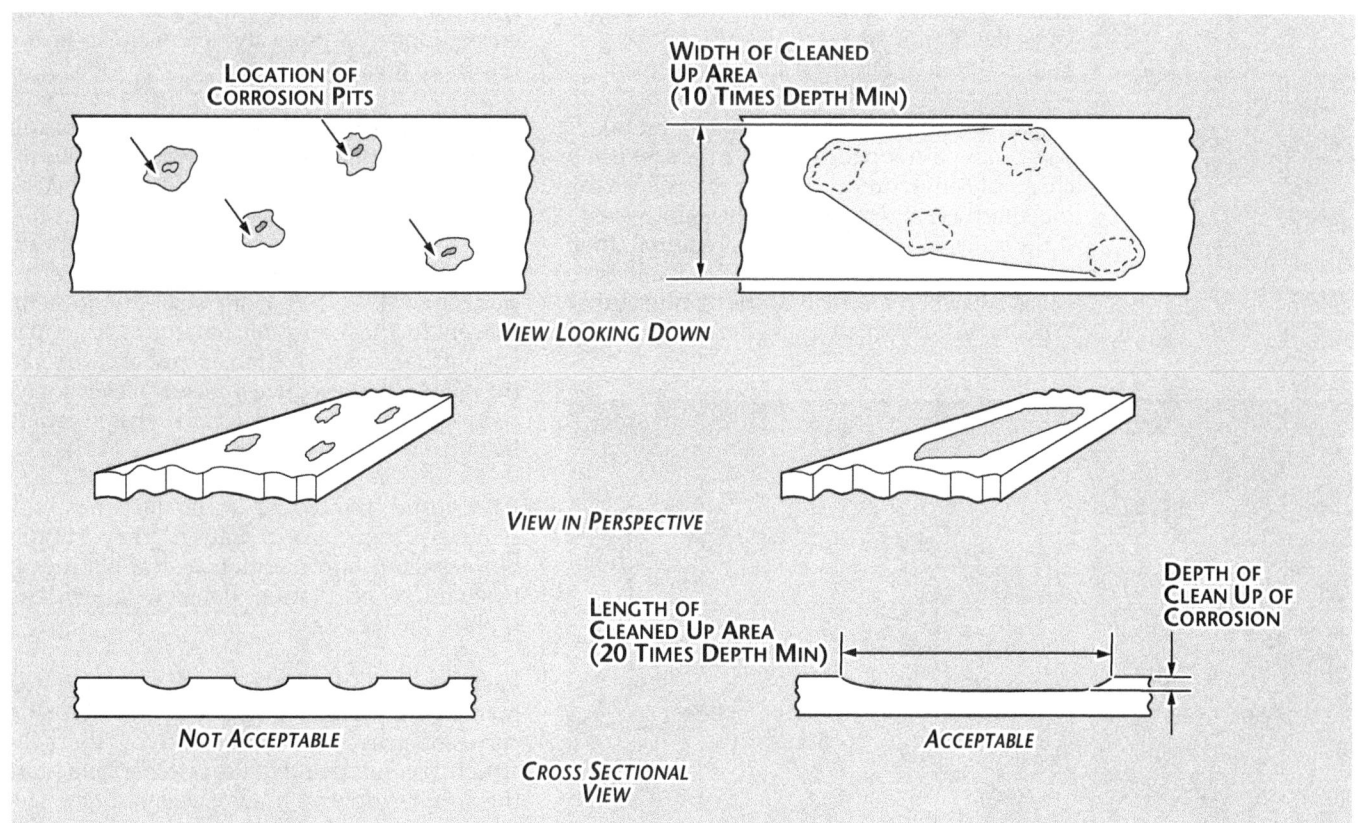

Figure 3-4-4. Example of acceptable clean-up of corrosion pits

sanding discs, and abrasive rubber mats are also available. The method used depends upon the metal and degree of corrosion.

Removal Methods

The removal method to use on each metal for each particular degree of corrosion is discussed below.

Aluminum and aluminum alloys. These are the most widely used materials for aircraft construction. Aluminum appears high in the electro-chemical series of elements and corrodes very easily. However, the formation of a tightly-adhering oxide film offers increased resistance under most corrosive conditions. Most metals in contact with aluminum form couples that undergo galvanic corrosion attack. The alloys of aluminum are subject to pitting, intergranular corrosion and intergranular stress corrosion cracking. In some cases the corrosion products of metal in contact with aluminum are corrosive to aluminum.

Special treatment of anodized surfaces. Anodizing is the most common surface treatment of aluminum alloy surfaces. The aluminum sheet or casting is the positive pole in an electrolyte bath in which chromic acid or other oxidizing agents produce a supplemental protective oxide film on the aluminum surface. The anodized surface coating offers the alloy a great deal of protection as long as it is not damaged. Once the film is damaged, it can only be partially restored by chemical surface treatment.

Repair of aluminum alloy sheet metal. If water can be trapped in blended areas, use a chemical conversion coating in accordance the manufacturer and reestablish the paint system. If the same level and contour is required, then the area may be filled with structural adhesive or sealant, when specified by the manufacturer. When areas are small enough that structural strength has not been significantly decreased, no other work is required prior to applying the protective finish.

When corrosion removal exceeds the limits of the structural repair manual, a DER or the aircraft manufacturer should be contacted for repair instructions.

Where exterior doublers are installed, it is necessary to seal and insulate them adequately to prevent further corrosion.

- Doublers should be made from Alclad metal or as specified, and the sheet should be anodized, or a chemical conversion coat applied, after all cutting, drilling, and countersinking has been accomplished.
- All rivet holes should be drilled, countersunk, surface treated, and primed before installation of the doubler.
- Suitable sealing compounds should be applied in the area to be covered by the doubler. Apply sufficient thickness of sealing compound to fill all voids in the repair area.

Install rivets wet with sealant. Sufficient sealant should be squeezed out into holes so that all fasteners, as well as all edges of the repair plate, will be sealed against moisture.

Remove all excess sealant after fasteners are installed. Apply a fillet sealant bead around the edge of the repair. After the sealant has cured apply the protective paint finish to the reworked area.

Corrosion removal around countersunk fasteners. Intergranular corrosion in aluminum alloys often originates at countersunk areas where steel fasteners are used. When corrosion is found around a fixed fastener head, the fastener must be removed to ensure corrosion removal. All corrosion must be removed to prevent further corrosion and loss of structural strength. To reduce the recurrence of corrosion, the panel should receive a chemical conversion coating, be primed, and have the fasteners installed wet with sealant.

Each time removable steel fasteners are removed from access panels, they should be inspected for condition of the plating. If mechanical or plating damage is evident, replace the fastener.

Fastener installation. Some of the suggested methods of installing fasteners are to brush a corrosion-preventive compound on the substructure around and in the fastener hole, start the fastener, apply a bead of sealant to the fastener countersink, set and torque the fastener within the working time of the sealant. If this

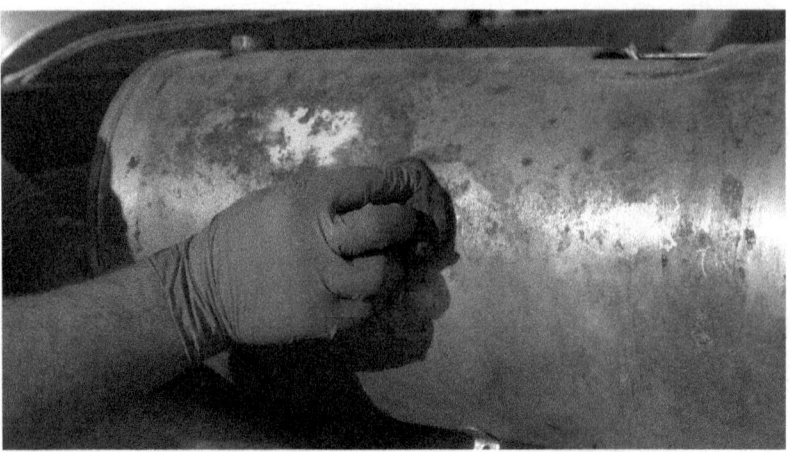

Figure 3-4-5. Light corrosion can be removed with aluminum wool.

Figure 3-4-6. Many chemicals are acidic; therefore proper precautions should be followed.

is not possible, apply the primer and the corrosion preventive compound to the substructure and fastener, set and torque the fastener while the primer is wet.

Light to moderate corrosion. The removal of corrosion products by hand can be accomplished by use of aluminum grit and silicon carbide abrasive, such as non-woven, non-metallic, abrasive mat, abrasive cloth, and paper. Aluminum wool, fiber bristle brushes, and pumice powder are also acceptable methods as is shown in Figure 3-4-5.

Stainless steel brushes may be used as long as the bristles do not exceed 0.010 inch in diameter. After use of this brush, the surface should be polished with 60 grit aluminum oxide abrasive paper, then with 400 grit aluminum oxide paper. Care should be exercised in any cleaning process to avoid breaking the protective film.

NOTE: *Steel wool, emery cloth, steel wire brushes (except stainless steel), copper alloy brushes, rotary wire brushes, or severe abrasive materials should not be used on any aluminum surface.*

Chemical Corrosion Removal

Corrosion removal compounds and pre treatments are acidic materials and may be used to remove corrosion products from aluminum alloy materials or skins, stringer, ribs in wings, tubing, or ducts in accordance with the manufacturer's recommendations.

Wear acid-resistant gloves, protective mask and protective clothing when working with these acid compounds. Apply the solution by flowing, mopping, sponging, brushing, or wiping. When applying the solution to large areas, begin the application at the lowest area and work upward, applying the solution with a circular motion to disturb the surface film and ensure proper coverage.

Chemical removal of corrosion begins with stripping the paint as shown in Figure 3-4-6.

CAUTION: *When working with acidic materials, keep the solution away from magnesium surfaces. The solution must be confined to the area being treated. All parts and assemblies including cadmium-plated items and hinges susceptible to damage from acid should be masked and/or protected. Also mask all openings leading to the primary structure that could trap the solution and doors or other openings that would allow the solution to get into the aircraft or equipment.*

Allow the solution to remain on the surface for several minutes then rinse away with clean water. For pitted or heavily-corroded areas the compound will be more effective if applied warm (140°F/60°C) followed by vigorous agitation with a non-metallic acid-resisting brush or aluminum oxide abrasive nylon mat. Allow sufficient dwell time before rinsing. After each application, examine the pits and/or corroded area with a 10-power magnifying glass to determine if another application is required. Corrosion that is still on the area will appear as a powdery crust slightly different in color from the uncorroded base metal. Darkening of area due to shadows and reaction from the acid remover should not be considered.

Once the corrosion has been removed and the area rinsed with clean water, a chromate conversion coating that meets MIL-C-81706, MIL-C-5541 or Alodine 1200, must be applied immediately thereafter.

Moderate to heavy corrosion. Where the problem is severe enough to warrant the use of power tools, either a pneumatic drill motor driving an aluminum oxide-impregnated nylon abrasive wheel, flap brush or rubber-grinding wheel with an abrasive value to approximately 120 grit is recommended. Corrosion-removal accessories, such as flap brushes or rotary files, should only be used on one type of metal. For example, do not use a flap brush used to remove aluminum to remove magnesium, steel, etc. Pneumatic sanders may be used with disk and paper acceptable for use on aluminum.

When mechanically removing corrosion from aluminum, especially aircraft skin, thinner than 0.0625 inch, extreme care must be used.

Vigorous, heavy, continuous abrasive grinding can generate enough heat to cause metallurgical change. If heat damage is suspected, hardness tests or conductivity tests must be accomplished to verify condition of the metal. The use of powered rotary files should be limited to heavy corrosion and not on skin thinner than 0.0625 inch.

Blasting. Abrasive blasting may be used on aluminum alloys using glass or grain abrasive as an alternate method of removing corrosion from clad and non-clad aluminum alloys. This method is shown in Figure 3-4-7.

Abrasive blasting should not be used to remove heavy corrosion products. Direct pressure machines should have the nozzle pressure set at 30 to 40 p.s.i. for clad aluminum alloys and 40 to 45 p.s.i. for non-clad aluminum alloys. Engineering approval should be obtained prior to abrasive blasting metal thinner than 0.0625 inch. When using abrasive blasting on aluminum alloys, do not allow the blast stream to dwell on the same spot longer than 15 seconds. Longer dwell times will cause excessive metal removal. Intergranular exfoliation corrosion is not to be removed by abrasive blasting; however, blasting may be used with powered corrosion removal to determine whether all exfoliation corrosion has been removed. Inspect the area for remaining corrosion. Repeat procedure if any corrosion remains.

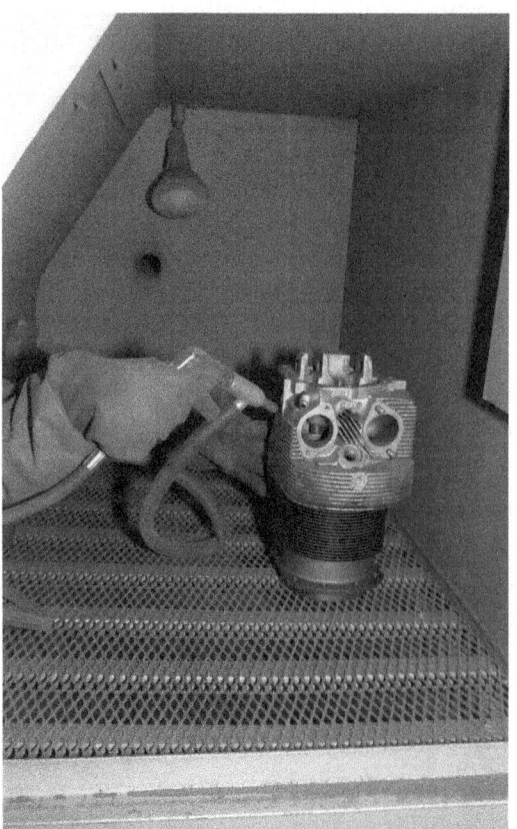

Figure 3-4-7. Glass beading aluminum castings

NOTE: *If corrosion remains after the second attempt, use a stronger method, e.g., chemical to mechanical.*

Using a blend ratio of 20:1 (length to depth) blend and finish the corrosion rework area with progressively finer abrasive paper until 400-grit paper is used.

Clean reworked area using dry cleaning solvent. Determine depth of faired depressions to ensure that rework limits have not been exceeded. Figure 3-4-8 details this method.

Apply a chemical conversion coating immediately after reworking. If 48 hours or more have elapsed since the conversion coating was first applied and the primer or final paint system has not yet been applied, then reapply the conversion coating before continuing. Apply paint finish to area.

CAUTION: *These solutions should not be allowed to come in contact with magnesium or high-strength steels (180,000 p.s.i.). Do not permit solutions or materials to contact paint thinner, acetone or other combustible material as fire may result.*

Magnesium and magnesium alloys. These are the most chemically active of the metals used in aircraft construction and are the most difficult to protect. However, corrosion on magnesium surfaces is probably the easiest to detect in its early stages. Since magnesium corrosion products occupy several times the volume of the original magnesium, initial signs show a lifting of the paint films and white spots on the magnesium surface. These rapidly develop into snow-like mounds or even white whiskers. The prompt and complete correction of the coating failure is imperative if serious structural damage is to be avoided.

Corrosion will usually occur around the edges of skin panels, underneath hold-down washers, or in areas physically damaged by shearing, drilling, abrasion, or impact. Entrapment of moisture under and behind skin crevices is frequently a contributing factor. If the skin section can be easily removed, this should be accomplished to ensure complete inhibition and treatment.

Treatment of wrought magnesium. When practical, complete mechanical removal of corrosion products on magnesium should be practiced. Mechanical cleaning should normally be limited to the use of stiff bristle brushes and similar nonmetallic cleaning tools.

Any entrapment of steel particles from steel wire brushes, steel tools, or contamination of treated surfaces, or dirty abrasives, can cause more trouble than the initial corrosive attack.

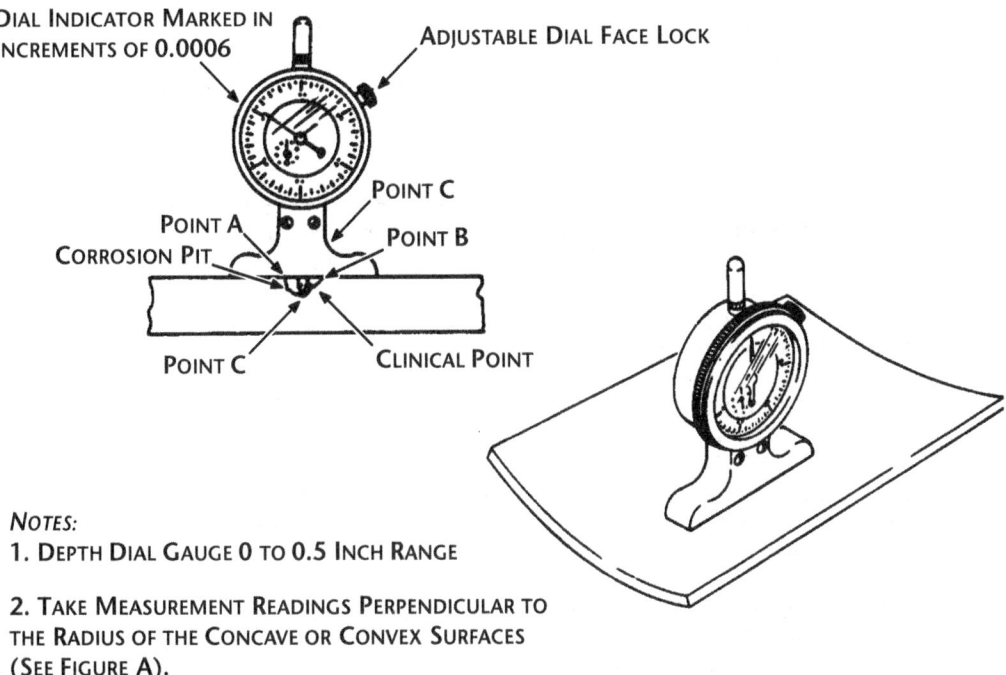

NOTES:
1. DEPTH DIAL GAUGE 0 TO 0.5 INCH RANGE

2. TAKE MEASUREMENT READINGS PERPENDICULAR TO THE RADIUS OF THE CONCAVE OR CONVEX SURFACES (SEE FIGURE A).

3. TAKE MEASUREMENT READINGS ON EDGES ADJACENT TO CORROSION DAMAGE OR BLENDED DEPRESSION (POINTS A AND B). TAKE SEVERAL READINGS AROUND CENTER OF CORROSION DAMAGE OR BLENDED DEPRESSION (SEE FIGURE A) AND CONSIDER THE DEEPEST READING TO BE POINT C. DETERMINE THE DEPTH AS FOLLOWS:

Figure 3-4-8. Measuring the depth of blended areas

The following procedural summary is recommended for preliminary treatment of corroded magnesium areas under most field conditions:

- When aluminum insulating washers are used and they no longer fasten tightly to magnesium panels, corrosion is likely to occur under the washers if corrective measures are not taken.

- When machine screw fasteners are used, aluminum insulating washers must be removed from all locations to surface treat the magnesium panel.

- Where permanent fasteners other than machine screws are used, the insulating washer and fastener must be removed.

- When located so water can be trapped in the counter-bored area where the washer was located, use sealants to fill the counterbore. If necessary, fill several areas adjacent to each other. It may be advantageous to cover the entire row of fasteners with a strip of sealant.

Repair of magnesium sheet metal. The same general instructions apply when making repairs in magnesium as in aluminum alloy skin, except that two coats of epoxy primer may be required on both the doubler and skin being patched instead of one. Where it is difficult to form magnesium alloys in the contour, aluminum alloy may be utilized. When this is done, it is necessary to ensure effective dissimilar metal insulation. Vinyl tape will ensure positive separation of dissimilar metals, but edges will still have to be sealed to prevent entrance of moisture between mating surfaces where repairs are made. It is recommended that only non-corrosive type sealant be used, since it serves a dual purpose of material separation and sealing.

In-place Treatment of Magnesium Castings

Magnesium castings are more porous and more prone to penetrating attack than wrought magnesium skin. However, treatment in the field is, for all practical purposes, the same for all magnesium. Bellcranks, fittings, and numerous covers, plates, and handles may also be magnesium castings. When attack occurs on a casting, the earliest practical treatment is required to prevent dangerous corrosive penetration.

Magnesium engine cases that are immersed in salt water can develop moth holes and complete penetration overnight.

If at all practical, faying surfaces involved shall be separated to treat the existing attack effectively and prevent its further progress. The same general treatment sequence as detailed for magnesium skin should be followed. Where engine cases are concerned, baked enamel overcoats are usually involved rather than other topcoat finishes. A good air-drying enamel can be used to restore protection.

If extensive removal of corrosion products from a structural casting is involved, a decision from the aircraft manufacturer or a DER may be necessary to evaluate the adequacy of structural strength remaining. Refer to the aircraft manufacturer if any questions of safety are involved.

Light to moderate corrosion. Non-powered removal can be accomplished using abrasive mats, cloth, and paper with aluminum oxide grit (do not use silicon carbide abrasive). Metallic wools and hand brushes compatible with magnesium such as stainless steel and aluminum, may be used. When a brush is used the bristles should not exceed 0.010 inch in diameter. After using a brush, the surface should be polished with 400 grit aluminum oxide abrasive paper, then with 600 grit aluminum oxide abrasive paper. Pumice powder may be used to remove stains or to remove corrosion on thin metal surfaces where minimum metal removal is allowed.

Chemical corrosion removal. Chemical corrosion removal on magnesium alloys is usually done with a chromic acid pickle solution. Chemical corrosion removal methods are not considered adequate for areas that have deep pitting, or heavy corrosion and corrosion by products. Areas that previously had corrosion removed by mechanical means, or have previously been sand blasted also are not suitable for chemical removal methods since too much material may have been removed.

> **NOTE:** *Do not use this method for parts containing copper and steel-based inserts (unless the inserts are masked off) and where it might come into contact with adhesive bonded skins or parts.*

A solution of chromium trioxide in water may be used to remove surface oxidation and light corrosion products from magnesium surfaces.

Mask off nearby operating mechanisms, cracks and plated steel to keep the solution from attacking them. Apply a chromic acid solution carefully to the corroded area with an acid-resistant brush. Allow the solution to remain on the surface for approximately 15 minutes, agitating as required. Thoroughly rinse the solution from the surface with plenty of clean water.

Repeat the sequence as necessary until all corrosion products have been removed and the metal is a bright metallic color.

Moderate to heavy corrosion. Powered corrosion removal can be accomplished using pneumatic drill motor with either an aluminum-oxide-impregnated abrasive wheel, flap brush, or rubber grinding wheel with an abrasive of 120 grit.

Also a rotary file with fine flutes can be used for severe or heavy corrosion product buildup on metals thicker than 0.0625 inch. If a flap brush or rotary file is used, it should only be used on one type of metal. Do not use either a hand or rotary carbon steel brush on magnesium.

Pneumatic sanders are acceptable if used with disk or paper of aluminum oxide. When using sanders, use extra care to avoid over heating aircraft skins thinner than 0.0625 inch. Do not use rotary wire brushes on magnesium.

> **CAUTION:** *Cuttings and small shavings from magnesium can ignite easily and are an extreme fire hazard. Fires of this metal must be extinguished with a Class D, metals, fire extinguisher.*

Blasting. Abrasive blasting is an approved method of corrosion removal on magnesium alloys of a thickness greater than 0.0625 inch. Remove heavy corrosion products by hand brushing with a stainless steel or fiber brush followed by vacuum abrasive blasting with glass beads or grain abrasive at an air pressure of 10 to 35 p.s.i. Upon completion of blasting, inspect for the presence of corrosion in the blast area. Give particular attention to areas where pitting has progressed into intergranular attack. This is necessary because abrasive blasting has a tendency to close up streaks of intergranular corrosion rather than remove them if the operator uses an improper impingement angle. If the corrosion has not been removed in a total blasting time of 60 seconds on any one specific area, other mechanical methods of removal should be utilized.

> **CAUTION:** *When blasting magnesium alloys, do not allow the blast stream to dwell on the same spot longer than 15 seconds. Longer dwell times will cause excessive metal removal.*

Inspect the reworked area to ensure that no corrosion products remain. If corrosion products are found, repeat the method used and re-inspect.

Depressions resulting from rework should be faired, using a blend ratio of 20:1. Clean rework

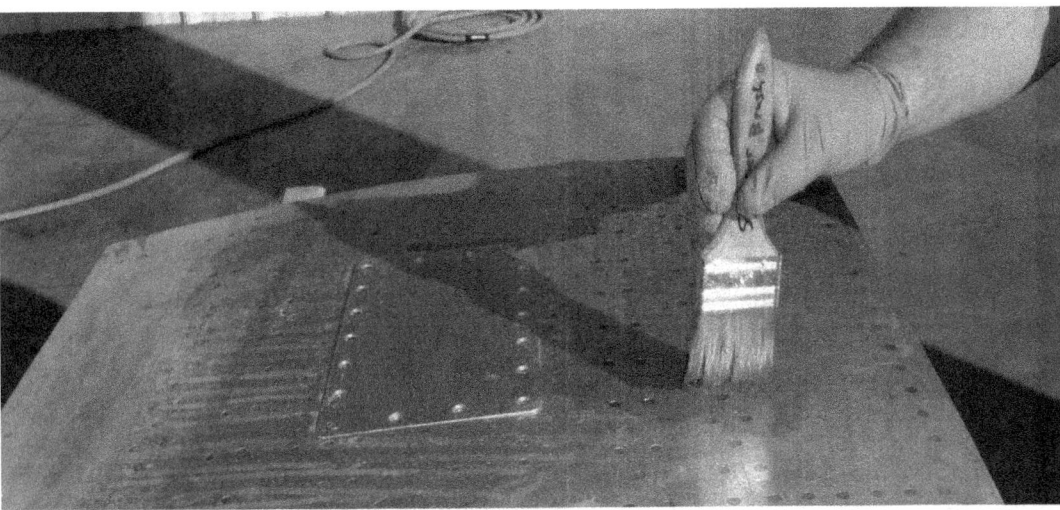

Figure 3-4-9. Applying chromic acid pretreatment to a skin

area using 240 grit abrasive paper. Smooth with 300 grit and finally polish with 400 grit abrasive paper.

Determine the depth of faired depressions to ensure that rework limits have not been exceeded. Refer to the manufacturer's specifications. Clean the reworked area using a solvent to provide a water-break-free surface.

Apply chromic acid pretreatment. Chemical pretreatment such as a chromic acid solution provides a passive surface layer with an inhibitive characteristic that resists corrosive attack and provides a bond for subsequent coatings. Properly applied magnesium pretreatments tend to neutralize corrosion media in contact with the surface.

Chromic acid pretreatment may be applied to all magnesium parts that require touch-up. This treatment is generally used in refinishing procedures or when parts and assemblies are too large to be immersed. This treatment is less critical to apply than the other brush-on treatments. It is relatively inexpensive and not as harmful when trapped in faying surfaces.

Apply the solution by brush, swab, or flow on using low-pressure spray (non-atomizing) until the metal surface becomes a dull color (the color can vary from green/brown, brassy, yellow-brown to dark-brown). For good paint adhesion, a dark-brown color free of powder is considered best. The color may vary in using different vendors' materials. Figure 3-4-9 shows a chromic acid treatment.

> **NOTE:** *Good application requires proper preparation of the chromic acid coating solution and cleaning of the surface where the solution will be applied. A water-break test is recommended if the cleanliness of the surface is in doubt. Too long an exposure to the brush-on solution produces coatings that will powder and impair adhesion of applied paint finish/films.*

Observe the coating closely for color changes during treatment, rinse with cold running water when the desired condition or color is reached, then air dry. Prepare and use test panels made of the same material and under the same conditions before starting the actual treating operation. This procedure may be used to determine the application time required to produce the required coating. A good coating is uniform in color and density, adheres well and is free of loose powder. Apply primer and topcoat finish. Remove masking and protective coverings.

Ferrous Metals

One of the most familiar kinds of corrosion is red iron rust. Red iron rust results from atmospheric oxidation of steel surfaces. Some metal oxides protect the underlying base metal, but red rust is not a protective coating. Its presence actually promotes additional attack by attracting moisture from the air and acts as a catalyst to promote additional corrosion.

Red rust first shows on bolt heads, hold down nuts, and other unprotected aircraft hardware. Red rust will often occur under nameplates on steel parts. Its presence in these areas is generally not dangerous. It has no immediate effect on the structural strength of any major components. However, it shows a general lack of maintenance and may indicate attack in more critical areas.

When paint failures occur or mechanical damage exposes highly stressed steel surfaces to the atmosphere, even the smallest amount of rusting is potentially dangerous and should be immediately removed.

Treatment of high-strength steel. High-strength steels are those that have been heat-treated above Rockwell C40, with 180,000 p.s.i. tensile strength.

Any corrosion on the surface of a highly stressed steel part is potentially dangerous, and the careful removal of corrosion products is mandatory. Surface scratches or change in surface structure from overheating can cause sudden failure of these parts.

Acceptable removal methods include careful use of mild abrasive mats, cloths, and papers such as fine grit aluminum oxide, metallic wool, or fine buffing compounds.

Undesirable methods include the use of any power tool because of the danger of local overheating and the formation of notches that could lead to failure. Only use the technique shown in Figure 3-4-10 with the manufacturer's approval.

The use of chemical corrosion removers is prohibited without engineering authorization because high-strength steel parts are subject to hydrogen embrittlement.

Treatment of Stainless Steel

Stainless steels are of two general types: magnetic and nonmagnetic.

Magnetic stainless steels. Magnetic steels are ferritic or martensitic and are identified by numbers in the 400-series. Corrosion often occurs on 400-series stainless steels, and treatment is the same as that specified for high-strength steels.

Nonmagnetic stainless steels. Nonmagnetic stainless steels are austenitic, identified by numbers in the 300-series. They are more corrosion resistant than the 400-series steels, particularly in a marine environment.

Figure 3-4-10. Powered removal must be done in accordance with the manufacturer's recommendations.

Austenitic steels develop corrosion resistance by an oxide film, which should not be removed even though the surface is discolored. The original oxide film is normally formed at time of fabrication by passivation. If this film is broken accidentally or by abrasion, it may not restore itself without repassivation.

If any deterioration or corrosion does occur on austenitic steels, and the structural integrity or serviceability of the part is affected, it will be necessary to remove the part.

Removing corrosion from ferrous metals. If possible, corroded steel parts should be removed from the aircraft. When impractical to remove the part mask the area to prevent affecting any surrounding structure.

> **NOTE:** *Use of acid-based strippers, chemical removers, or chemical conversion coatings are not permitted on steel parts without engineering authorization.*

Determine extent of corrosion damage and remove residual corrosion by hand sanding with mild abrasive mats, cloths, and papers, such as fine aluminum oxide grit. Heavy deposits of corrosion products may be removed by approved mechanical methods for that particular form of steel and/or stainless steel.

Inspect the area for remaining corrosion. Repeat the procedure if any corrosion remains, the structural integrity of the part is not in danger, and the part meets the rework limits established by the manufacturer or FAA authorized DER.

Generally, depressions should be faired using a blend ratio of 20:1. Clean the area using 240-grit paper. Smooth area with 300-grit paper and give final polish with 400-grit paper.

Determine that the depth of faired depression of the rework have not exceeded the limits. Clean the reworked area with dry cleaning solvent and apply a protective finish or a specific organic finish as required.

> **NOTE:** *Steel surfaces are highly reactive immediately following corrosion removal; consequently, primer coats should be applied within one hour after sanding.*

Other Metals and Alloys

Silver, platinum, and gold finishes are used in aircraft assemblies because of their resistance to ordinary surface attack and their improved electrical or heat conductivity. Silver-plated electrodes can be cleaned of brown or black sulfide tarnish, by placing them in contact with a piece of magnesium sheet stock while immersed in a warm water solution of common table salt mixed with baking soda or by using a fine grade abrasive mat or pencil eraser followed by solvent cleaning. If assemblies are involved, careful drying and complete displacement of water is necessary. In general, cleaning of gold or platinum coatings is not recommended in the field.

Copper and copper alloys. These are relatively corrosion resistant, and attack on such components will usually be limited to staining and tarnish. Such change in surface condition is not dangerous and should ordinarily have no effect on the function of the part. However, if it is necessary to remove such staining, a chromic acid solution of 8 to 24 ounces per gallon of water containing a small amount of battery electrolyte (not to exceed 50 drops per gallon) is an effective brightening bath. Immerse the stained part in the cold solution. Surfaces can be treated in place by applying the solution to the stained surface with a small brush.

Avoid any entrapment of the solution after treatment. Clean the part thoroughly following treatment to remove all residual solution.

Staining may also be removed using a fine grade abrasive mat followed by solvent cleaning.

Serious copper corrosion is evident by the accumulation of green-to-blue copper salts on the corroded part. Remove these products mechanically using a stiff bristle brush, brass wire brush, 400-grit abrasive paper or bead blast with glass beads. Air pressure when blasting should be 20 to 30 p.s.i. for direct pressure machines. Do not bead blast braided copper flexible lines. Reapply a surface coating over the reworked area. Chromic acid treatment will tend to remove the residual corrosion products.

> **CAUTION:** *Brushing, sanding, and abrasive blasting of copper and copper alloys can be dangerous due to the creation of toxic airborne particles. Take necessary precautions to ensure safety.*

Titanium and titanium alloys. Titanium and its alloys are highly corrosion-resistant because an oxide film forms on their surfaces upon contact with air.

When titanium is heated, different oxides having different colors form on the surface. A blue oxide coating will form at 700 to 800°F; a purple oxide at 800 to 950°F; and a gray or black oxide at 1000°F or higher. These coatings are protective discolorations and should not be removed.

Corrosive attack on titanium surfaces is difficult to detect. It may show deterioration from the presence of salt deposits and metal impurities at elevated temperatures so periodic removal of surface deposits is required. However, if corrosion develops on titanium, it usually occurs as pitting. Acceptable methods for corrosion removal are the use of stainless steel wool or hand brush.

Abrasive mats, cloths, and papers with either aluminum oxide or silicon carbide grit may be used as well.

Dry abrasive blasting using glass beads or aluminum oxide at a blast pressure of 40 to 50 p.s.i. followed with hand polishing with aluminum polish and soft cloth is acceptable.

> **CAUTION:** *Dry abrasive blasting of titanium alloys creates sparking. Ensure that hazardous concentrations of flammable vapors are not present.*

> **NOTE:** *Titanium surfaces are susceptible to hydrogen embrittlement that can induce stress corrosion and associated pitting. Therefore, chemicals such as fire-resistant hydraulic fluids must be controlled. Chlorinated hydrocarbon solvents and chemical corrosion removers are prohibited from use on titanium and titanium alloys.*

Plated parts. Chromium- and nickel- plated parts are used extensively as protective and wear-resistant coatings over high-strength steel parts (landing gear journals, shock strut pistons, etc.). Chromium and nickel plate provide protection by forming an impervious physical coat over the underlying base metal. When breaks occur in the surface, the protection is destroyed.

The amount of reworking that can be performed on chromium and nickel-plated components is limited. The rework should consist of light buffing to remove corrosion products and produce the required smoothness. The buffing should not take the plating below the minimum allowable thickness.

Whenever a chromium or nickel-plated component requires buffing, coat the area with a corrosion-preventive compound, if possible.

When buffing exceeds the minimum thickness of the plating, or the base metal has sustained corrosive attack, the component should be removed and replaced.

The removed component can be restored to serviceable condition by having the old plating completely stripped and re-plated in accordance with acceptable methods and specifications.

Cadmium- and zinc-plate. Cadmium plating is used extensively in aircraft construction as a protective finish over both steel and copper alloys.

Protection is provided by a sacrificial process in which the cadmium is attacked rather than the underlying base material. Properly functioning cadmium surface coatings may show mottling, ranging from white to brown to black spots on their surfaces. These show the sacrificial protection being offered by the cadmium coat, and under no condition should such spotting be removed merely for appearances' sake. In fact, cadmium will continue to protect even when actual breaks in the coating develop and bare steel or exposed copper surfaces appear.

When the breakdown of the cadmium plating occurs and the initial appearance of corrosion products on the base metal develops, some mechanical cleaning of the area may be necessary but should be limited to removal of the corrosion products from the underlying base material. Under no condition should such a coating be cleaned with a wire brush. If protection is needed, a touch-up with primer or a temporary preservative coating should be applied. Restoration of the plate coating cannot be done in the field.

Zinc coatings offer protection in an identical manner to cadmium, and the corrective treatment for failure is generally the same as for cadmium-plated parts. However, the amount of zinc on aircraft structures is very limited and usually does not present a maintenance problem.

Section 5

Special Conditions for Corrosion Control

Corrosion-proofing of Land Planes Converted to Sea Planes

A special problem is encountered in the conversion of land planes to seaplanes. In general, land planes do not receive corrosion proofing to the same extent as seaplanes. Corrosion-proofing standards for land planes converted to seaplanes are divided into two classes: necessary minimum precautions and recommended precautions. A converted land plane is shown in Figure 3-5-1.

Regardless of such precautions, it is imperative that the exterior surfaces of seaplanes be washed with clear fresh water immediately following extended water operation, or at least once a day when operated in salty or brackish water. Wash interior surfaces of seaplanes exposed to spray, taking care to prevent damage to electrical circuits or other items subject to injury.

Minimum precautions. The following procedures are considered the minimum to safeguard the airworthiness of converted aircraft and are not in themselves intended to maintain airworthiness for an indefinite period.

Unless already protected, treat exposed fittings or fittings that can be reached through inspection openings with two coats of zinc chromate primer, Paralketone, non-water soluble heavy grease, or comparable materials. This applies to items such as wing-root fittings, wing-strut fittings, control-surface hinges, horns, mating edges of fittings, and attached bolts.

Coat non-stainless control cables with grease or Paralketone or other comparable protective coating, if not replaced with corrosion resistant cables.

Inspect all accessible sections of the aircraft structure. Clean structural parts showing corrosion and refinish if the corrosion attack is superficial. If a part is severely corroded, replace it with an adequately corrosion-proofed part.

Recommended precautions. These precautions are suggested as a means of maintaining an aircraft in condition for safe operation over extended periods.

To assist in the detection of corrosion, additional inspection openings in the aircraft are recommended. Experience has shown that openings to allow inspection of the lower and rearward portion of the fuselage are particularly desirable.

Additional provisions should be made for free drainage and ventilation of all interiors to prevent collection of moisture by using scoop-type marine drain grommets.

The interior of structural steel tubing can be protected by watertight sealing or by flushing with hot linseed oil and plugging all openings. All tubing must be inspected for missing sealing screws and the presence of entrapped water. Corrosion around sealing screws, at welded clusters and bolted fittings may be indicative of entrapped moisture.

The fabric on the bottom of fabric-covered aircraft should be slit longitudinally on the bottom of the fuselage and tail structure for access to these sections. Coat the lower structural members with zinc chromate primer (two coats); followed by a coat of dope-proof paint or cellophane tape and repair the fabric. This precaution is advisable within a few months after start of operation as a seaplane.

The interior of metal-covered wings and fuselages should be sprayed with an adherent corrosion inhibitor.

Prevent the entry of water by sealing, as completely as possible, all openings in wings, fuselage, control-surface members, openings for control cables, and tail-wheel wells.

Aircraft Recovered from Water Immersion

Aircraft recovered from partial or total immersion in standing water or flash floods require an in-depth inspection and cleaning of both the exterior and interior areas. Water-immersion increases the probability of corrosive attack, it removes lubricants, deteriorates aircraft materials, and destroys electrical and avionics components.

Sea water, because of its salt content, is more corrosive than fresh water. However, fresh water may also contain varying amounts of salt and as drying occurs the salt concentration is increased and corrosive attack accelerated.

Prompt action is the most important factor following recovery of an aircraft from water-immersion. Components of the aircraft which have been immersed, such as the power plant, accessories, airframe sections, actuating mechanisms, screws, bearings, working surfaces, fuel and oil systems, wiring, radios, and radar should be disassembled as necessary and the contaminants completely removed.

Initial wash. As soon as possible after the aircraft is recovered from water immersion, thoroughly wash all internal and external areas of the aircraft using a water detergent solution as specified by the aircraft manufacturer.

If the recommended cleaning compound is not available, use any available mild household detergent solution with fresh tap water.

Reciprocating engines and propellers. The propeller should be removed from the engine and the engine from the aircraft. The exterior of the engine and propeller should be washed with steam, or fresh hot water.

Accessories. Major accessories and engine parts should be removed and all surfaces flushed with fresh hot water. If facilities are available, immerse the removed parts, time permitting, in hot water or hot oil, for a short time. Soft water is preferred. Change the water frequently. All parts must be completely dried by air blast or other means. If no heat-drying facility is available, wipe the cleaned parts with suitable drying cloths.

Propellers. The constant-speed propeller mechanism should be disassembled, as required, to permit complete decontamination. Clean all the parts with steam or fresh hot water. Dry the cleaned parts in an oven, but if a heat drying facility is not available, wipe the cleaned parts with suitable drying cloths.

Airframe. The salvageable components of the fuselage, wings, empennage, hulls and floats, and movable surfaces should be processed in the following manner. The fabric from fabric-covered surfaces should be removed and replaced. Clean the aircraft interior and exterior using steam under pressure with a steam cleaning compound. Direct the steam into all seams and crevices where corrosive water may have penetrated. Avoid steam cleaning electrical equipment, such as terminal boards and relays. Those areas that have been steam cleaned should be rinsed immediately with either hot or cold fresh water.

Touch-up all scratches and scars on painted surfaces using zinc chromate primer or preservative. Un-drained hollow spaces or fluid entrapment areas should be provided temporary draining facilities by drilling out rivets at the lowest point. Install new rivets after drainage.

Remove and replace all leather, fabric upholstery, and insulation. Plastic or rubber foam that cannot be cleaned of all corrosive water must be replaced.

All drain plugs or drive screws in tubular structures should be removed and the structure blown out with compressed air. If water has reached the tubular interiors, carefully flush with hot fresh water and blow out water with compressed air. Roll the structure as necessary to remove water from pockets. Fill the tubes with hot linseed oil, approximately 180°F, drain oil and replace drain plugs or drive screws.

Clean sealed wood and non-metallic areas (excluding acrylic plastics), with warm water. Replace wood, and other porous materials exposed to water immersion unless surfaces are adequately sealed to prevent penetration by water. Virtually all solvents and phenolic type cleaning agents are detrimental to acrylics and will either soften the plastic or cause crazing.

Remove instruments, radios, applicable wiring harnesses and plumbing, and repair and inspect as necessary.

Figure 3-5-1. A typical land plane to seaplane conversion

Photo courtesy of Cessna Aircraft

Chapter 4
Aging aircraft

Section 1
Background on Aging Aircraft

On April 28, 1988, multiple fatigue cracks caused an Aloha Airlines Boeing 737 to lose part of its upper fuselage. Although the aircraft was able to land safely, the accident resulted in the death of one flight attendant and injuries to many passengers. The aircraft, which entered service in April of 1969, had accumulated 35,496 hours and 89,690 flight cycles. Though durability and damage tolerance were issues before the time of the accident, the Aloha accident is generally considered the beginning of the FAA's focused Aging Aircraft Program. Figure 4-1-1 shows the Aloha Airlines B-737 after landing.

The first response to the accident was an industry-wide review of aircraft design and maintenance programs. The aviation community discovered that with proper maintenance and structural modifications as well as attention to service-related damage such as fatigue and corrosion, the service lives of airplanes could be safely extended. Airworthiness Directives (ADs) were issued to ensure that susceptible structures would not degrade below acceptable limits. These and other ADs also ensured attention to proper maintenance and inspection.

To address mid- and long-range issues, the industry established the Aging Aircraft Task Force (now known as the Airworthiness Assurance Working Group, or AAWG).

A 1993 report by the AAWG identified 14 multi-site damage (MSD) and multi-element damage (MED) structural details to be evaluated and monitored. Though the lap splice implicated in the Aloha accident would remain the top prior-

Learning Objectives:

- Regulations for Aging Aircraft
- Structural Integrity Program
- Widespread Fatigue Damage
- Damage Tolerance
- Best Practices

Left: **Age and usage are two factors that affect aging aircraft.**

ity for investigation, there were now potential widespread fatigue damage (WFD) threats to occupy engineers and scientists. In 1999, another AAWG report added additional structural details to the list of those of earlier concern. That report contained industry recommendations for addressing the problems detailed in the report.

Though the Aloha accident was the impetus for the FAA's Aging Aircraft Program, the opportunity to identify and address potential problems before they could emerge as threats to aviation safety was apparent. The FAA concluded that improvements to the management of large commercial transports should be leveraged to initiate similar improvements in the management of commuter aircraft and helicopters. Neither class of aircraft utilized the more sophisticated damage tolerance concept for design and maintenance.

Not every safety issue received appropriate attention before another catastrophic accident occurred. In 1989, the fan disk on the number two engine of a DC 10 disintegrated, resulting in the loss of all hydraulics. The accident resulted in the establishment of the Titanium Rotating Components Review Team and the Engine Titanium Consortium (ETC).

In the summer of 1996, a Boeing 747 exploded over the Atlantic shortly after take off from New York's John F. Kennedy Airport. The National Transportation Safety Board determined that the probable cause of the accident was an explosion of the center wing fuel tank resulting from ignition of the flammable fuel-air mixture in the tank. The source of ignition could not be determined with certainty; but of the sources evaluated by the investigation, the most likely one was a short circuit outside of the tank that allowed excessive voltage to enter it though electrical wiring associated with the fuel quantity indication system. A schematic of that system is shown in Figure 4-1-2.

The AAWG has recommended that any aging aircraft inspection program consist of a records review and the development of a Supplemental Structural Inspection Document (SSID). From the SSID, the Supplemental Structural Inspection Program (SSIP) is derived. When the two documents are completed, the certificate holder may develop an inspection program for airplanes that are in service.

Because of the accidents mentioned above, the safety of nonstructural systems has become a major priority of the National Aging Aircraft Research Program.

Structural Integrity Challenges

Structural problems in aging aircraft are particularly difficult to detect because the damage in question consists of multiple interacting flaws in areas that are difficult to access and inspect. To find and address the problem, the development of sophisticated inspection methods are required. These methods must allow for multiple interacting cracks and their effect on the structure of the aircraft.

Damage tolerance methods. Fail-safe design allows a structure to retain its required residual strength for a period of use after the failure of a principal structural element. A key element of fail-safe design is damage tolerance. This is the characteristic of a structure or component that allows it to retain its required residual strength for a period of use after the item has sustained a given level of corrosion, fatigue or accidental damage.

Damage tolerance-based inspections are developed by a manufacturer or operator based on an engineering evaluation of likely sites where damage could occur. They are also developed in consideration with expected stress levels,

Figure 4-1-1. The structural failure of the upper cabin of an Aloha Airlines 737-200 was the event that put national focus on aging aircraft and corrosion inspections.

Photo courtesy of NTSB

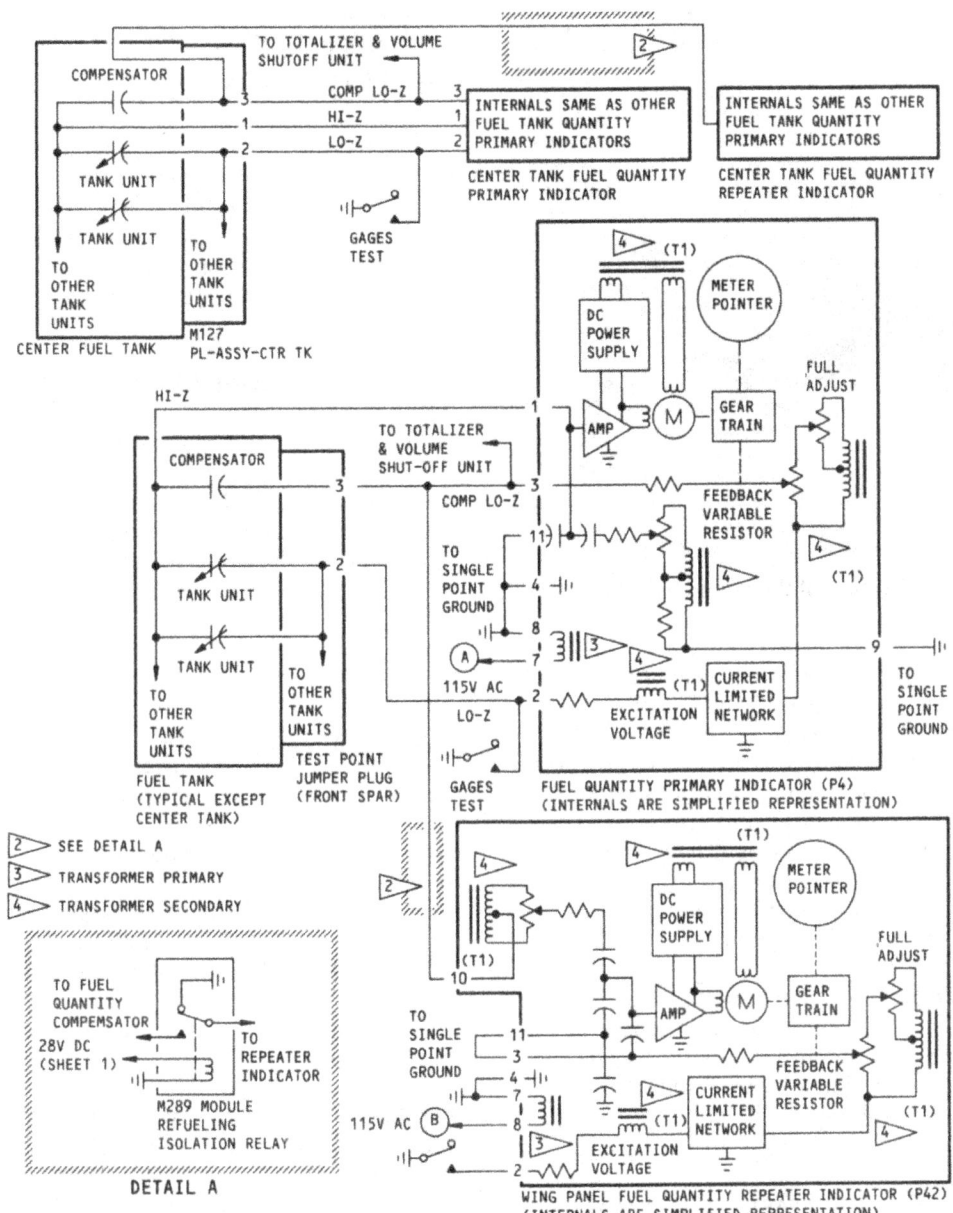

Figure 4-1-2. Schematic of the B747 fuel quantity system

material characteristics and projected damage growth rates. It is important to understand that the damage tolerance concept allows an aircraft to continue in operation with a known level of deterioration. By acknowledging that damage is present, how quickly it is likely to spread and the fail-safe nature of its design, the operator can schedule the repair at the lowest possible cost without compromising safety.

Future challenges. The continued challenges to the aviation industry involve many areas of science and engineering. The development of advanced technological programs creates problems in the management and implementation of the developed methods. The challenge is to focus on the problems and not on the technology used to solve the problem. Some areas of concern are:

- Aging commercial fleet
- Widespread Fatigue Damage (WFD)
- Regulatory requirements

Aging commercial fleet. Aircraft were designed to be inspected visually. Inspection methods and intervals are established during certification and through service experience. As the aircraft ages, these inspections become increasingly problematic. In Figure 4-1-3, a 40-year-old B727 is still in service.

Areas of the aircraft that are inaccessible, multiple failure modes and unique structures all complicate the process of inspection. Many inspections will be even more difficult if the repeat interval forces operators to perform them at other than a regularly scheduled interval.

Figure 4-1-3. Older aircraft are still a vital part of the commercial fleet.

Widespread fatigue damage (WFD). Aircraft susceptible to widespread fatigue damage pose a substantial threat to safety. The identification of the conditions leading to WFD is not fully documented; researchers continue to examine the causes and effects. It is fairly clear that for an equivalent level of risk, critical crack lengths can be larger for structures with discrete damage than for structures with multi-site or element damage.

Regulatory requirements. The Federal Aviation Regulations are precise regarding the detection of damage on aircraft. They may not allow for timely application of improved inspection techniques.

ITEMS	FAR PART 121 OPERATIONS	FAR PART 135 OPERATIONS
Total years in service of the airplane	X	X
Total time in service of the airframe	X	X
Total flight cycles of the airframe	X	
Date of last inspection and records review	X	X
Current status of life limited parts of the airframe	X	X
Time since last overhaul of time specific structural items	X	X
Current inspection status of airplane	X	X
Current status of ADs	X	X
List of major structural alterations	X	X
Report of major structural repairs and their inspections	X	X

Table 4-1-1. Required records for FAA review

Federal Aviation Regulation for Aging Aircraft

A synopsis of the FARs is provided here in order to help with understanding the regulations. As with all regulatory material, only the most current information should be used. The FAA maintains the most current revisions to the regulations on its website at www.faa.gov.

FAR part 121 and 135 Aircraft

For aircraft and operators under FAR parts 121 and 135, the regulations contain the information that applies to all airplanes that are operated by a certificate holder, except for those airplanes operated between any points within the State of Alaska. Operators under FAR part 135 include those with multiengine airplanes with nine or fewer seats that are used in a scheduled operation.

Inspection and records review. After December 8, 2003, a certificate holder, foreign carrier or person may not operate an airplane under this part unless the FAA has notified the certificate holder that they have completed the required aging airplane inspection and records review. During the inspection and records review, the certificate holder must show the FAA that the maintenance of age-sensitive parts and components of the airplane has been adequate and timely enough to ensure the highest degree of safety.

Airplane and records availability. The certificate holder must make available to the FAA each airplane for which an inspection and records review is required. The airplane must be in a condition for inspection as specified by the FAA. The records must contain the information listed in Table 4-1-1.

The airplane shown in Figure 4-1-4 is typical of airplanes that are used in FAR part 121 operations.

A certificate holder, foreign carrier or person may not operate a transport category, turbine powered airplane with a type certificate issued after January 1, 1958, that has been increased in capacity to:

- A passenger seating capacity of 30 or more
- A maximum payload capacity of 7,500 pounds or more

NOTE: *Airplanes operated by a certificate holder between any points within the State of Alaska are exempt.*

After December 20, 2010, a certificate holder, foreign carrier or person, may not operate an

Figure 4-1-4. An airplane used in FAR part 121 operations.

Figure 4-1-5. This is the type of airplane operated under FAR part 135 regulations.

airplane unless the following requirements have been met:

1. The maintenance program for the airplane includes FAA-approved damage-tolerance-based inspections and procedures for structure that is susceptible to fatigue cracking that could contribute to a catastrophic failure. These inspections and procedures must take into account the adverse affects repairs, alterations, and modifications may have on fatigue cracking and the inspection of the airplane structure.

2. The damage-tolerance-based inspections and procedures, and any revisions to these inspections and procedures, must be approved by the Aircraft Certification Office or office of the Transport Airplane Directorate with oversight responsibility for the relevant type certificate or supplemental type certificate, as determined by the Administrator. The certificate holder must include the damage-tolerance-based inspections and procedures in the certificate holder's FAA-approved maintenance program.

In Figure 4-1-5, a typical aircraft that is operated in a FAR part 135 operations is shown.

Section 2

A Structural Integrity Program

Service experience has shown there is a need to have continuing updated knowledge on the structural integrity of transport airplanes, especially as they age. The structural integrity of these airplanes is of concern because such factors as fatigue cracking and corrosion are time-dependent and our knowledge about them can best be assessed based on real-time operational experience and the use of the most modern tools of analysis and testing.

The Federal Aviation Administration, type certificate holders, and operators have continually worked to maintain the structural integrity of older airplanes. Traditionally, this has been carried out through an exchange of field service information, subsequent changes to inspection programs and by the development and installation of modifications on particular aircraft.

Increased use, longer operational lives, and safety demands imposed on the current fleet of transport airplanes indicate the need for a program that ensures a high level of structural integrity for all airplanes in the transport fleet. Accordingly, inspection and evaluation programs are intended to ensure a continuing structural integrity assessment by each airplane manufacturer, and the incorporation of the results of each assessment into the maintenance program of each operator.

The basis for a maintenance and inspection program is found in two documents; the Supplemental Structural Inspection Program (SSIP) and the Supplemental Structural Inspection Document (SSID). Only after these documents and their programs have been approved can an inspection program be initiated.

Supplemental Structural Inspection Programs

The type certificate holder (TCH), in conjunction with operators, is expected to initiate the development of a Supplemental Structural Inspection Program (SSIP) for each airplane model. Such a program must be implemented before analysis, tests, and/or service experience indicates that a significant increase in inspection and/or modification is necessary to maintain structural integrity of the airplane. In the absence of other data as a guideline, the program should be initiated no later than the time when the high-time or high-cycle airplane in the fleet reaches one half its design service goal. This should ensure that an acceptable program is available to the operators when needed.

The program should include procedures for obtaining service information, assessment of service information, available test data, and new analysis and test data. A Supplemental Structural Inspection Document (SSID) should be developed, from this body of data. The type

Figure 4-2-1. Cracks which exceed the limits found during an inspection.

of damage that an SSIP can address is shown in Figure 4-2-1.

The recommended SSIP as well as the criteria used and the basis for those criteria should be submitted to the cognizant FAA Aircraft Certification Office for review and approval. The SSIP should be adequately defined in the SSID. The SSID should include the type of damage being considered, likely sites and inspection access. The damage threshold, interval, method and procedures as well as the applicable modification status and/or life limitation, and the types of operations for which the SSID is valid are considered part of the document.

FAA review. The FAA's review of the SSID will include both engineering and maintenance aspects of the proposal. Because the SSID is applicable to all operators and is intended to address potential safety concerns on older airplanes, the FAA will make it mandatory under the existing Airworthiness Directive (AD) system. In addition, the FAA will issue ADs to implement any service bulletins or other service information publications found to be essential for safety during the initial assessment process. Service bulletins or other service information publications revised or issued as a result of in-service findings resulting from implementation of the SSID should be added to the SSID or they will be implemented by separate AD action as appropriate.

In the event an acceptable SSID cannot be obtained on a timely basis, the FAA may impose service life, operational, or inspection limitations to assure structural integrity.

The TCH should revise the SSID whenever additional information shows a need. The original SSID will normally be based on predictions or assumptions (from analyses, tests, and/or service experience) of failure modes, time to initial damage, frequency of damage, typically detectable damage, and the damage growth period. Consequently, a change in these factors sufficient to justify a revision would have to be substantiated by test data or additional service information. Any revision to SSID criteria and the basis for these revisions should be submitted to the FAA for review and approval of both engineering and maintenance aspects.

Evaluation for Widespread Fatigue Damage

The likelihood of fatigue damage occurring in an airplane's structure increases with an airplane's usage. The design process generally establishes a design service goal (DSG) in terms of flight cycles/hours for the airframe. It is expected that any cracking that occurs on an airplane operated up to the DSG will occur in isolation (i.e., local cracking), originating from a single source, such as a random manufacturing flaw like a poorly drilled fastener hole or a localized design detail. It is considered unlikely that cracks from manufacturing flaws or localized design issues will interact strongly as they grow. The SSIP are intended to find this form of damage before it becomes critical.

With extended usage, a uniformly loaded structure may develop cracks in adjacent fastener holes or in adjacent similar structural details. These cracks, while they may or may not interact, can have an adverse effect on the large damage capability (LDC) before the cracks become detectable. The development of cracks at multiple locations may also result in strong interactions that can affect subsequent crack growth, in which case the predictions for local cracking would no longer apply. An example of this situation may occur at any skin joint where load transfer occurs and is shown in Figure 4-2-2. Simultaneous cracking at many fasteners along a common rivet line may reduce the residual strength of the joint below required levels before the cracks are detectable under the routine maintenance program established at time of certification.

The TCH, in conjunction with operators and in some cases the operators themselves, are expected to initiate development of a maintenance program with the intent of precluding operation with WFD. Such a program must be implemented before analysis, tests, and/or service experience indicates that widespread fatigue damage may develop in the fleet. To ensure that an acceptable program is available when needed, development of the program should be initiated no later than the time when the highest-time or highest-cycle airplane in the fleet reaches three quarters of its designed service goal (DSG) or the extended service goal (ESG).

The results of the WFD evaluation should be presented for review and approval to the cognizant FAA Aircraft Certification Office having type certificate responsibility for the airplane model being considered. It is expected that the TCH will work closely with operators in the development of these programs to assure that the expertise and resources are available when implemented.

The FAA review. The FAA's review of the WFD evaluation results includes both engineering and maintenance aspects of the proposal. Since

Figure 4-2-2. A skin joint can be a source of cracks on any aircraft.

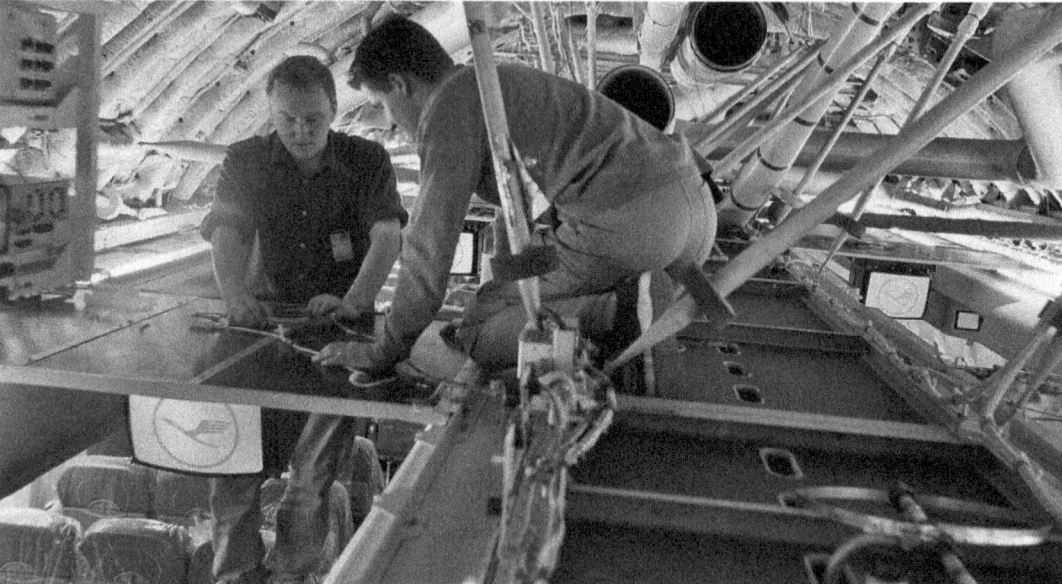

Figure 4-3-1. A view of the interior of a B747 in a D level inspection. *Photo courtesy of Lufthansa*

WFD is a safety concern for all operators of older airplanes, identified inspection or modification and/or replacement programs are proposed to be made mandatory by operational rules applicable to 14 CFR parts 91, 121, 125, 129, and 135. In addition, any service bulletins or other service information publications revised or issued because of in-service findings resulting from implementation of these programs may require separate AD action.

In the event an acceptable WFD evaluation is not completed in a timely manner, the FAA may impose service-life restrictions, operational limitations, or inspection requirements to ensure structural integrity.

It is expected that the original recommended actions stemming from a WFD evaluation will be focused on those structural items that are soon expected to reach a point at which multiple site damage (MSD) or multiple element damage (MED) is predicted to occur. As the fleet ages, more areas of the airplane will reach the age at which MSD/MED is predicted to occur and the recommended service actions should be updated accordingly. New service experience findings, improvements in the prediction methodology, better load spectrum data, or a change in any of the factors upon which the WFD evaluation is based may dictate a revision to the evaluation.

Operators will be expected to accomplish a WFD evaluation of applicable modified, repaired, or altered structure. The results must be presented for review and approval to the cognizant FAA Aircraft Certification Office having type certificate responsibility for the airplane model being considered.

Implementation. Once the FAA issues a SSID AD, operators must amend their current structural inspection programs to comply with and account for the applicable AD. The program to preclude WFD in the fleet has been mandated by operational rules which require operators to amend the current structural maintenance programs. Any ADs issued as a result of a WFD finding that require structural modification will be handled separately. In all cases, compliance is required in accordance with the applicable regulations.

Section 3

The Supplemental Structural Inspection Document

It is essential to identify the structural parts and components that contribute significantly to carrying flight, ground, pressure, or control loads, and whose failure could affect the structural integrity necessary for the continued safe operation of the airplane. The damage-tolerance or safe-life characteristics of these parts and components must be established or confirmed.

Analyses made in respect to the continuing assessment of structural integrity should be based on supporting evidence, including test and service data. This supporting evidence should include consideration of the structural loading distributions, and material behavior. An inspection threshold may be based solely on a statistical assessment of fleet experience, if it can be shown that equal confidence can be placed in such an approach.

An effective method of evaluating the structural condition of older airplanes is selective inspection with intensive use of non-destructive techniques. Another effective method is inspecting individual airplanes by partial or complete dismantling ("teardown") of available structure. The interior of a B747 is shown undergoing an extensive inspection in Figure 4-3-1.

The effect of repairs or modifications approved by the manufacturer should be considered. In addition, it may be necessary to consider the effect of repairs and operator-approved modifications on individual airplanes. The operator has the responsibility for ensuring notification and consideration of any such aspects.

Damage-tolerant Structures

The damage-tolerance assessment of the airplane structure should be based on the best information available. The assessment should include a review of analysis, test data, operational experience, and any special inspections related to the type design. A determination should then be made of the site or sites within each structural part or component considered likely to crack, and the time or number of flights at which this might occur.

The growth characteristics of damage and interactive effects on adjacent parts in promoting more rapid or extensive damage should be determined. This determination should be based on study of those sites that may be subject to the possibility of crack initiation due to fatigue, corrosion, stress corrosion, disbonding, accidental damage, or manufacturing defects in those areas shown to be vulnerable by service experience or design judgment.

The minimum size of damage that is practical to detect and the proposed method of inspection should be determined. This determination should take into account the number of flights required for the crack to grow from detectable to the allowable limit, such that the structure has a residual strength corresponding to the conditions stated under the certification document.

> **NOTE:** *In determining the proposed method of inspection, consideration should be given to visual inspection, nondestructive testing, and analysis of data from built-in load and defect monitoring devices.*

The continuing assessment of structural integrity may involve more extensive damage than might have been considered in the original fail-safe evaluation of the airplane, such as:

- A number of small adjacent cracks, each of which may be less than the typically detectable length, developing suddenly into a long crack
- Failures or partial failures in other locations following an initial failure due to redistribution of loading causing a more rapid spread of fatigue
- Concurrent failure or partial failure of multiple load path elements (e.g., lugs, planks, or crack arrest features) working at similar stress levels

An allowable limit of the size of damage should be determined for each site such that the structure has a residual strength for the load conditions specified in the certification documents. The size of damage that is practical to detect by the proposed method of inspection should be determined, along with the number of flights required for the crack to grow from detectable to the allowable limit.

Inspection Program

The purpose of a continuing airworthiness assessment in the most basic terms is to adjust the current maintenance inspection program to assure continued safety.

The recommended inspection program should consider the following:

- Fleet experience, including all of the scheduled maintenance checks
- Confidence in the proposed inspection technique
- The joint probability of reaching the load levels described above and the final size of damage in those instances where probabilistic methods can be used with acceptable confidence

Inspection thresholds. These should be established for supplemental inspections. These inspections would be supplemental to the normal inspections and include detailed internal inspections.

For structure with reported cracking, the threshold for inspection should be determined by analysis of the service data and available test data for each individual case.

For structure with no reported cracking, it may be acceptable, provided sufficient fleet experience is available, to determine the inspection threshold on the basis of analysis of existing fleet data alone. This threshold should be set such as to include the inspection of a sufficient number of high-time airplanes to develop

Figure 4-3-2. The installation of a cargo door in this airplane is a major modification

Figure 4-3-3. This hush kit installation is a major alteration.

added confidence in the integrity of the structure. Thereafter, if no cracks are found, the inspection threshold may be increased progressively by successive inspection intervals until cracks are found.

Structural Modifications, Repairs and Alterations

All major modifications (STC's), repairs, and alterations that create, modify, or affect structure that are susceptible to MSD/MED (as identified by the TCH) must be evaluated to demonstrate the same confidence level as the original manufactured structure. The operator is responsible for ensuring the accomplishment of this evaluation. The operator may first need to conduct an assessment on each of its airplanes to determine what modifications, repairs, or alterations would be susceptible to MSD/MED. The following are some examples of types of modifications, repairs, and alterations that present such concerns:

- Passenger-to-freighter conversions, including addition of main deck cargo doors shown in Figure 4-3-2
- Gross weight increases, increased operating weights, increased zero fuel weights, increased landing weights and increased maximum takeoff weights
- Installation of fuselage cutouts, passenger entry doors, emergency exit doors or crew escape hatches, fuselage access doors and cabin window relocations
- Complete re-engine and/or pylon modifications
- Engine hush-kits and nacelle alterations of the kind shown in Figure 4-3-3
- Wing modifications, such as the installation of winglets or changes in flight control settings
- Alteration of wing trailing edge structure
- Modified, repaired, or replaced skin splice
- Any modification, repair, or alteration that affects several frame bays
- A modification that covers structure requiring periodic inspection by the operator's maintenance program
- A modification that results in operational mission change that significantly changes manufacturers load/stress spectrum (for example, a passenger-to-freighter conversion)
- A modification that changes areas of the fuselage from being inspected externally, using visual means, to not being able to be inspected (for example, a large external fuselage doubler that resulted in hidden details, rendering them hidden visually) as can be seen in Figure 4-3-4

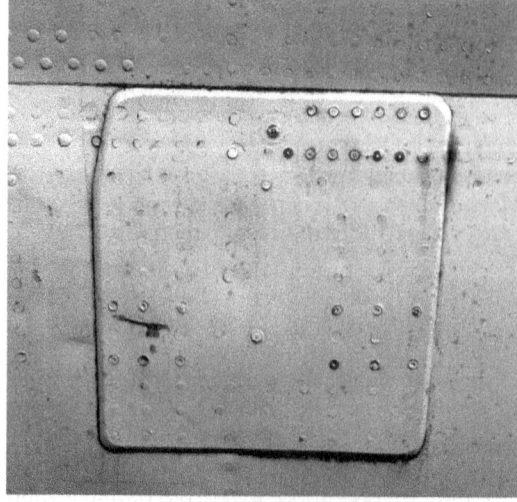

Figure 4-3-4 This external doubler hides the original structure from inspection.

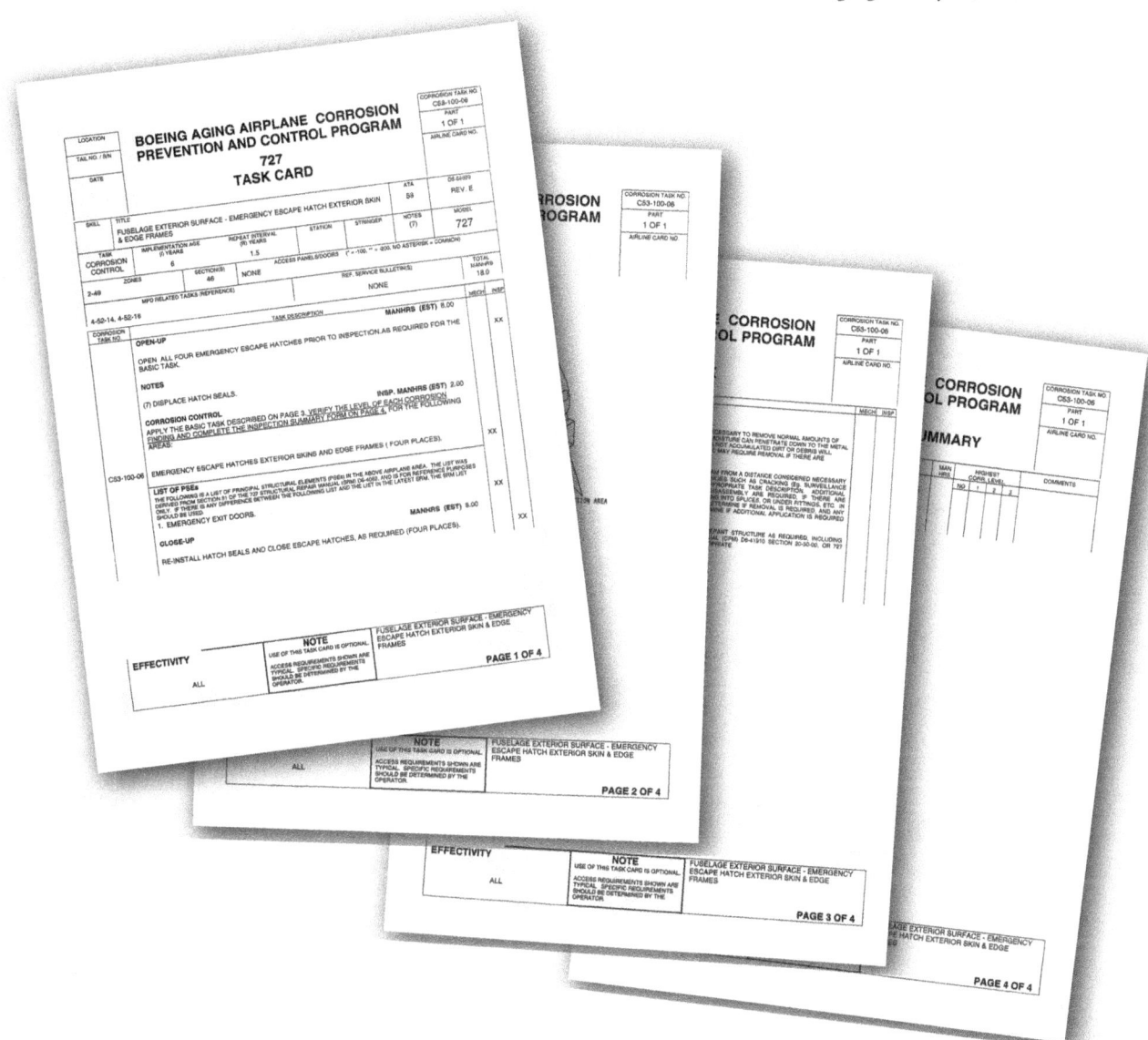

Figure 4-3-5. Inspection task cards for a Boeing 727

Inspection Program for Transport Aircraft

The actual inspection programs are developed after the SSID and the SSIP are approved and implemented. The SSID and SSIP form the foundation of the steps that are needed in the inspection program. The inspection areas, time intervals, methods and repairs are based on the service experience for the fleet of aircraft that are used by the operator. Figure 4-3-5 shows examples of inspection cards for a Boeing 727.

Figure 4-3-6 shows excerpts from the Corrosion Protection manual for the same aircraft. These pages contain expanded information for the technician who is inspecting the area detailed on the inspection card. Both of these documents are used together. Without the task cards the inspection steps cannot be performed. Without the information from the manual proper corrosion removal and preservation cannot be conducted.

Section 4

Inspection Programs for General Aviation Aircraft

The GA fleet is still in service well beyond the flight hours and years envisioned when the airplanes were first designed. In 2000, the average age of the nation's 150,000 single-engine fleet was more than 30 years. By 2020, the average age could approach 50 years.

There is concern that continued airworthiness and safety will become a bigger problem as the fleet ages. Several factors are key in keeping the existing fleet in service.

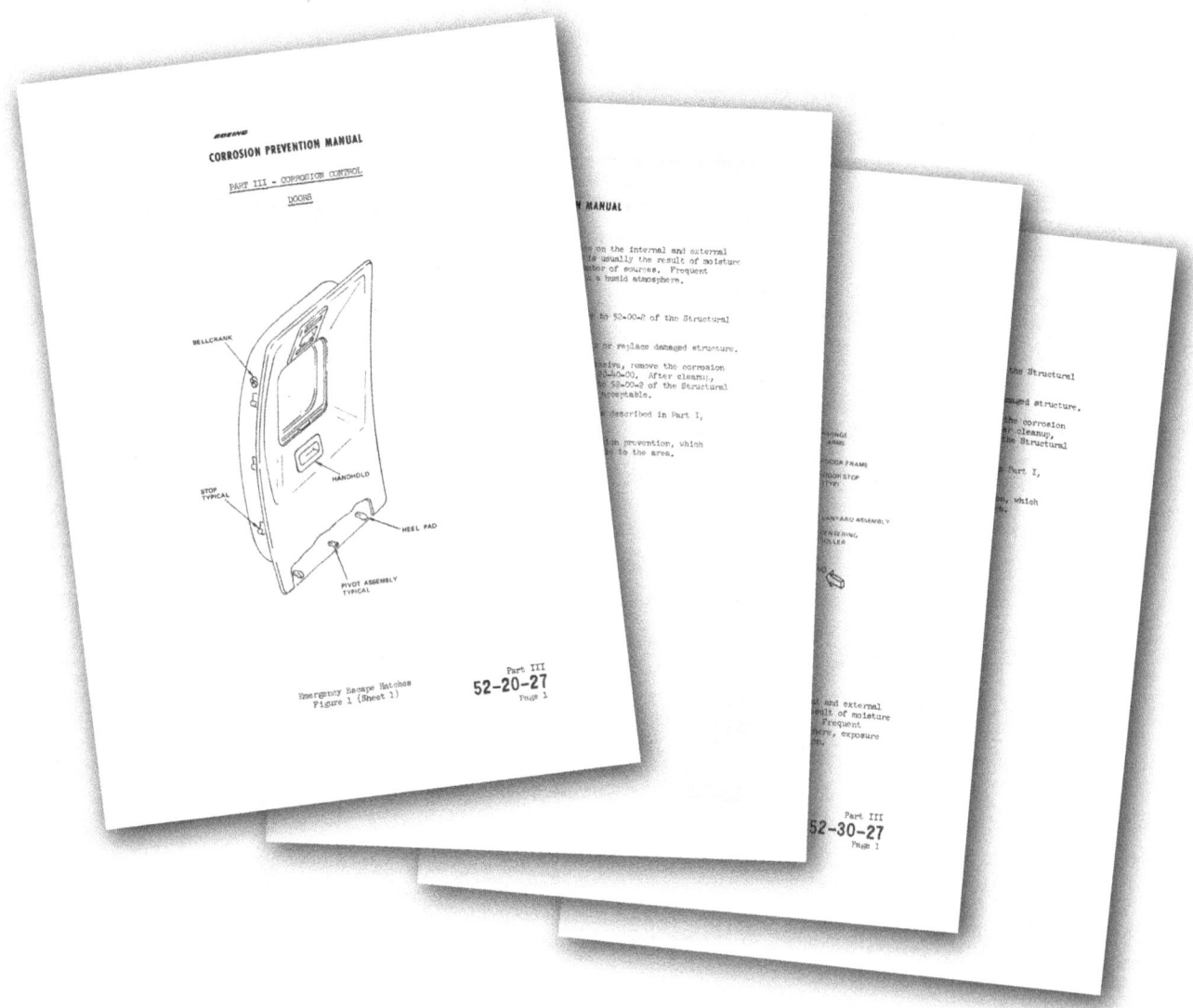

Figure 4-3-6. The pages from the MMM contain information on correct procedures

Studies show that the biggest safety concern is pilot situational awareness. To address this concern, modifications to the cockpit are now popular because of the rapid advances in avionics and associated affordability. A 40-year-old, four-place airplane with new avionics can remain productive for many years with periodic avionics upgrades as this technology advances. These improvements will increase safety and allow much of the existing fleet to remain in service well into the present century. At the same time, these airplanes could develop serious age-related problems as they continue to be used well beyond their envisioned design life. Figure 4-4-1 is a Cessna 173 that is 40 years old. It is capable of being flown daily.

The bulk of the fleet is designed to Civil Aviation Regulations (CAR) 3 standards that were established in the 1950s or earlier. These standards lack fatigue and continued airworthiness requirements as part of their certification. Thanks to the robust designs, these airplanes show few signs that they are aging aircraft. However, little is known about the condition of these older airplanes and how they change with age.

Industry and government have worked together to learn more about the effects of aging on aircraft. The focus of their efforts has been on research and regulation that specifically applies to large transports and commercial operations. However, the physics of aging ignore regulatory boundaries. Through this effort, industry has learned much about corrosion, metal fatigue, inspection techniques, and wiring deterioration. Much of this knowledge can benefit GA.

Methods for mitigating the effects of airplane aging parallel those used in medicine. Advances in medical science continue to result in new methods of detecting precursors to serious health problems. Health professionals recommend increasingly more intrusive "inspections" as people age. People accept these recommendations and generally request more thorough physicals as they get older. Conversely, most small airplanes, regardless of age, are rarely (if ever) inspected beyond a non-intrusive annual or 100-hour inspections as required by 14 CFR 43.15, Appendix D.

The recommended practices discussed in this section are similar to suggested physical exams that doctors recommend. Each airplane ages differently depending on how it is maintained and used over its life. However, airplane design concepts are similar from model to model and from manufacturer to manufacturer.

Best Practices

Many aging aircraft designs that are still capable of safe and useful operation in today's environment have manufacturers that have gone out of business. Other manufacturers may still exist but do not have the capability of providing field support for the aging models. Engineering drawings, maintenance procedures, and other technical data are not available from these manufacturers.

Acquiring, organizing, preserving, and making available for easy access all available data greatly increases the likelihood for improvements in the maintenance and safe operation of a particular airplane. Extended to a model type or several model types, these actions can have an enormous impact on the future safe operation of the aging small airplane fleet.

Two specific best practices can have a fundamental impact on the way maintenance and inspection is approached for aging airplanes. These are:

- Airplane records research
- Special attention inspections

Doing either of these helps assess the condition of an airplane. Doing both is needed to thoroughly assess the effects of aging on an airplane and provide a method of monitoring its condition as it continues to age.

Airplane records research. Record research is the first step in determining the condition of an aging airplane. The degree of inspection necessary, as well as the determination of what items may have already been inspected, will come from thorough records research. This research will not only identify certain maintenance and usage characteristics of a particular airplane, it will also expose potential areas of attention pertinent to a model type or usage class.

Typically, inspection and overhaul recommendations contained in older GA airplane maintenance instructions do not provide adequate guidance regarding aging maintenance issues. Therefore, assessing the quality of maintenance during an airplane's life is important to determine what parts were replaced, if corrosion was ever a problem, and other maintenance factors that could lead to a concern about aging. This helps establish a baseline to determine what maintenance, repairs and/or alterations have been done and how well the airplane has been cared for.

Research from general model-type issues can be compared with individual airplane information to identify similarities and differences. This helps answer the question: "Does what I am seeing on this particular airplane match the history of the airplane and the type according to the available records?"

Following is a list of sources that should be use to determine both individual airplane and model type histories:

- Logbook entries
- Aircraft records
- Type Certificate Data Sheets (TCDS)
- Airworthiness Directives (ADs)
- Special Airworthiness Information Bulletins (SAIBs)
- Service Difficulty Reports (SDR)
- National Transportation Safety Board (NTSB)
- Supplemental Type Certificates (STC)
- General aviation airworthiness alerts

Logbook entries. This is traditionally where most owners/mechanics begin their investigation. Having logbooks that are complete back to when the airplane came off the production line is a plus, but for various reasons (loss, theft, destruction, etc.) these records are not always available. This is why acquisition of the records for the air-

Figure 4-4-1. A classic example of a 40-year old airplane that has many hours of useful life left.

Photo courtesy of Cessna Aircraft

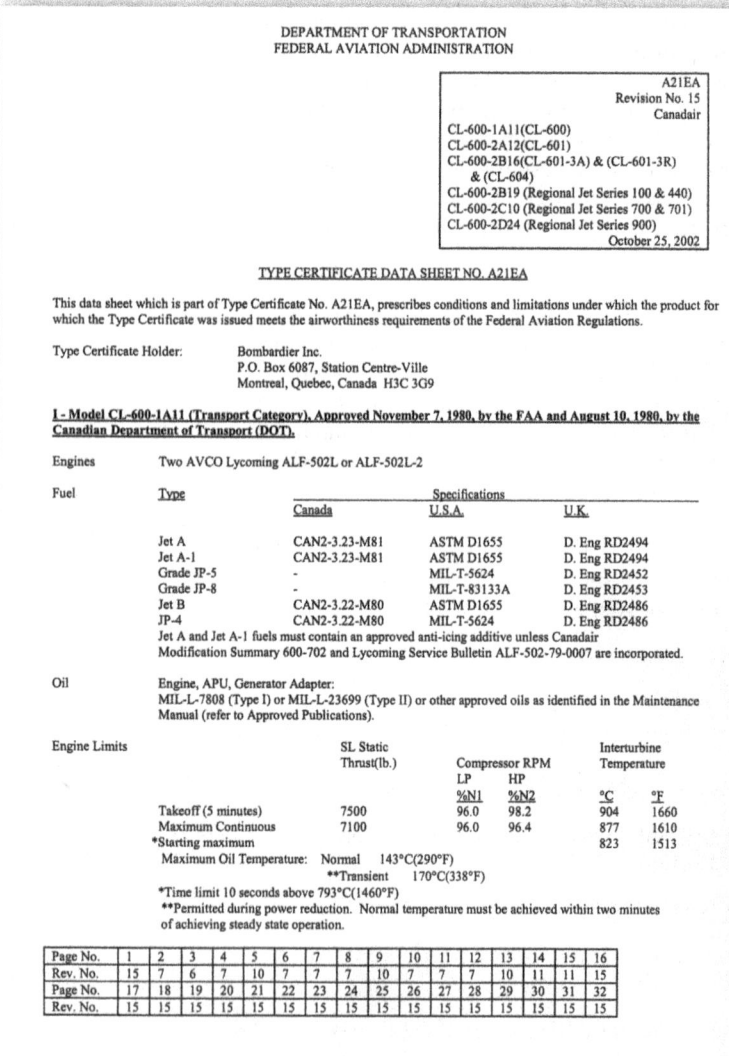

Figure 4-4-2. The front page of a 32 page TCDS. No aircraft may deviate from its type certificate without paperwork authorizing that deviation.

Other than the aircraft logbooks, the FAA maintains a database of information that can be accessed or requested on the internet. The primary sites are: www.faa.gov and www.av-info.faa.gov.

Many aircraft owners have formed model clubs that serve as clearinghouses for information on maintenance-related topics. Many of the clubs also have an internet presence with searchable databases. An internet search will reveal many of the locations for the information. All available resources should be used when investigating aging aircraft issues.

Aircraft records. The FAA provides records for specific aircraft, propellers, and engines. By sending a request to the FAA with the aircraft N number, an owner can receive a copy of all records pertaining that particular aircraft. Because airplane N numbers can change, a request for records by aircraft make, model, and serial number should be included. If these two do not match, then there may be a need to investigate further.

Available records include registrations, bills-of-sale, repair and alteration form 337s, supplemental type certificates (STC), and other information. Compare this paperwork to the physical airplane to determine if any unrecorded modifications have been made.

These records are provided either on microfiche, CD, or on paper for some older aircraft. There is a nominal charge to open the aircraft file and copy the material.

Type Certificate Data Sheets (TCDS). The TCDS, shown in Figure 4-4-2, contains relevant data specific to a model type. The FAA awards a type certificate (TC) only after the applicant shows compliance with the safety regulations as specified by the certification basis listed on the TCDS. The TCDS is a summary of the baseline technical description of the model type, which includes information such as performance, weights, center of gravity limits, and engine and propeller specifications.

Airworthiness Directives (AD). The FAA issues ADs to owners of affected airplane model types, engines, propellers, or appliances such as instruments. An AD is a mandatory action used to correct an unsafe condition. The required actions are usually modifications, one-time inspections, repetitive inspections, or a combination of the above. Owners should compare ADs for their airplane model type with their logbook entries to ensure that the ADs have been done. The first page of an AD for a general aviation airplane is shown in Figure 4-4-3.

plane from the FAA is important. The logbooks should show a clear trend of what maintenance has been preformed throughout the life of the airplane as well as the usage history of the airplane. There should also be clear indications of AD compliance as well as what modifications or major repairs have been carried out.

The logbook entries should be compared to the physical condition of the airplane. Always ask the question: "Does the logbook reflect what has actually been done to the airplane?" If so, then the owner should have confidence that the logbooks are reasonably accurate. If the logbook contains maintenance or alteration actions that are currently not part of the airplane, then further investigation may be in order to determine the importance of the missing action. Likewise, if the airplane is altered without any logbook entries, then the alterations should be investigated to determine the effect on the performance of the airplane.

Special Airworthiness Information Bulletins (SAIB). The FAA issues SAIBs to owners of affected airplane model types, engines, propellers, or appliances such as instruments. An SAIB is not mandatory but provides information regarding an airworthiness concern that is less serious than an unsafe condition that is addressed by an AD. SAIBs often reference manufacturer service bulletins and service letters.

Service bulletins/letters. Service bulletins and service letters are issued by the aircraft manufacturers to address in-service issues or for product improvement. These are often instructions for accomplishing the mandatory actions of an AD. Service bulletins and service letters can be obtained from the manufacturer or from model clubs. When a service letter or bulletin is part of an AD, compliance is mandatory.

Service Difficulty Reports (SDR). The FAA's database of SDRs contains reported maintenance and/or service problems for any aircraft, engine, or component. An airplane owner can search this database for model specific or individual airplane reports. This can be helpful for identifying areas that may be candidates for special attention (especially if the logbooks are incomplete). This can be useful for determining any past difficulties for specific airplanes.

In order to make the database as useful as possible, owners or mechanics are encouraged to submit SDRs with complete confidence that they will not lead to enforcement actions. Submitting an SDR can help alert other owners, mechanics, or inspectors of problems that may arise in aircraft no longer supported by an active manufacturer.

National Transportation Safety Board (NTSB). The NTSB keeps descriptions of more than 140,000 aviation accidents in a searchable database found at www.ntsb.gov. Searching this database helps the owners/operators or mechanics determine whether their particular aircraft has ever been involved in an accident. This could help the owner/operator or mechanic match up repairs on an aircraft to the reasons for those repairs. It can also help determine trends of accidents regarding a particular model of aircraft.

Supplemental Type Certificates (STC). Supplemental Type Certificates have been developed for many different types of aircraft. These are incorporated to upgrade or improve avionics, systems, engines, gross weight, etc. Design upgrades often have a positive effect with regard to aging issues. A review of the FAA STC database for a specific make and model may reveal design improvements to

[Federal Register: August 18, 2006 (Volume 71, Number 160)]
[Rules and Regulations]
[Page 47711-47714]
From the Federal Register Online via GPO Access [wais.access.gpo.gov]
[DOCID:fr18au06-6]

DEPARTMENT OF TRANSPORTATION

Federal Aviation Administration

14 CFR Part 39

[Docket No. FAA-2006-25262; Directorate Identifier 2006-CE-39-AD; Amendment 39-14725; AD 2006-17-04]

RIN 2120-AA64

Airworthiness Directives; The Cessna Aircraft Company Models 172R, 172S, 182T, T182T, 206H, and T206H Airplanes

AGENCY: Federal Aviation Administration (FAA), DOT.

ACTION: Final rule; request for comments.

SUMMARY: The FAA is adopting a new airworthiness directive (AD) for certain The Cessna Aircraft Company (Cessna) Models 172R, 172S, 182T, T182T, 206H, and T206H airplanes. This AD requires you to inspect the two end fittings on each of the flexible fuel hoses located in the engine compartment for the correct torque values, and, if any incorrect torque values are found during the inspection, tighten the hose end fittings to the correct torque values. This AD results from one report of loose fuel hose connections to the fuel injector servo on a Cessna Model 172S airplane. We are issuing this AD to detect and correct any incorrect torque values of the end fittings of flexible fuel hoses in the engine compartment, which could result in the loss of fuel flow and fuel leakage. Loss of fuel flow could result in partial or complete loss of engine power and fuel leakage could result in an engine compartment fire.

DATES: This AD becomes effective on September 1, 2006.

As of September 1, 2006, the Director of the Federal Register approved the incorporation by reference of certain publications listed in the regulation.

We must receive any comments on this AD by October 17, 2006.

ADDRESSES: Use one of the following addresses to comment on this AD.

DOT Docket Web site: Go to http:// dms.dot.gov and follow the instructions for sending your comments electronically.

Government-wide rulemaking Web site: Go to http://www.regulations.gov and follow the instructions for sending your comments electronically.

Mail: Docket Management Facility; U.S. Department of Transportation, 400 Seventh Street, SW., Nassif Building, Room PL-401, Washington, DC 20590-0001.

Fax: (202) 493-2251.

Hand Delivery: Room PL-401 on the plaza level of the Nassif Building, 400 Seventh Street, SW., Washington, DC, between 9 a.m. and 5 p.m., Monday through Friday, except Federal holidays.

Figure 4-4-3. The first page of an Airworthiness Directive for a General Aviation airplane.

address an aging issue discovered during routine inspection and maintenance.

General aviation airworthiness alerts. Also known as AC 43-16A, aviation maintenance alerts are compilations of recent maintenance problems that are showing up on aircraft (including factory- and amateur-built aircraft, helicopters, powerplants, and propellers), but have not been fully evaluated to the point of becoming a service bulletin, AD, or SAIB. The FAA publishes these alerts monthly from SDRs submitted by those who operate and maintain the civilian aircraft fleet. For aging aircraft, they can be an early indication of a developing trend or problem.

During the research described in the Logbook Entries and Aircraft Records sections, owners may discover that STCs are installed on their airplane. In some cases, an STC can affect the continued airworthiness of the airplane. If it does, this should be reflected with changes to the airplane's Instructions for Continued Airworthiness (ICA). Unfortunately, this does not always happen. Therefore, it is always in the owner's best interest to consider additional maintenance or inspection requirements for incorporated STCs that do not provide specific instructions for continued airworthiness.

During the record research, the owner should be aware of certain factors that may have significant impact on the condition of an aging airplane.

An airplane spends far more time on the ground than it does in the air. Therefore, the environment it is exposed to while on the ground plays a significant role in how it ages.

There should be adjustments to the way an individual airplane is inspected according to what the research reveals about the airplane's history. The research should answer several questions.

- Has the airplane been hangared?
- How much of its time has been outside?
- Where has the airplane been geographically?
- Has the airplane been inactive or in storage for a long period of time?
- Has the airplane been used in a special usage role?

If the airplane has spent much of its time outside, then there may be additional wear on seals, hoses, and moving parts exposed to the extremes of temperature and moisture. The likelihood of corrosion would be higher for an airplane not hangared.

Geographic Location

If it has been located in coastal areas, even for a few years, corrosion is a concern. Refer to Figure 3-1-2 in Chapter 3 for maps that show the areas of the world that are considered heavy corrosion risks. Corrosion degradation is not limited to structure; it can also cause problems with electrical connectors, instruments and engines. Corrosion is more of a concern if the airplane has not been hangared. If it has spent most of its time outside in areas where the temperature gets very cold or very hot, this will take an additional toll. Deterioration of electrical components, hoses, seals, and lubrication is faster when subjected to temperature extremes.

Inactive storage. Airplane inactivity has a more severe impact than regular use. The issues of material deterioration, lubrication, and part wear due to lack of movement can lead to accelerated aging. In many cases, airplanes are not properly prepared for long-term storage. The engine is not sealed, the airframe is open to rodents and birds and the application of sealants is overlooked.

Special use. A significant amount of time flying at low levels (for example, pipeline patrol or aerial survey) exposes the airframe to more frequent and higher gust loads. This in turn causes additional metal fatigue damage to the wings, empennage, and associated structure. Mountain flying is also a harsher gust environment and therefore more damaging. Operating the airplane with consistently heavy loads or for very short flights also induces additional fatigue damage. A Cessna 207 that is used in pipeline patrol is shown in Figure 4-4-4. Any operation in aerobatic or high-g maneuvers is damaging.

Unfortunately, metal fatigue damage does not heal over time. Severe usage early in an airplane's life is just as damaging as similar usage to an old airplane. Just because the airplane was new when it flew in severe operations, unseen and undetected damage to the metal still occurred and will remain. This can manifest itself in a higher likelihood of cracking later in the airplane's life.

Figure 4-4-4. Special uses for aircraft can stress the structures and create hidden problems. *Photo courtesy of Cessna Aircraft*

COMPONENT	SERVICE LETTERS AND/OR ADVISORY MATERIAL	DATE LAST INSPECTED/REPLACED	FINDINGS/ NOTES
GENERAL			
Weigh aircraft for actual weight and balance report			
Special attention for corrosion of magnesium parts (i.e., control surfaces, castings, etc.)			
AVIONICS			
Antenna (requires removal)			
Connectors			
Static dischargers			
Bonding			
ELT			
CONTROLS			
Control surface			
Inaccessible areas of control surfaces			
Attach points			
Inaccessible control cable runs			
Control surface balance weights			
ELECTRICAL			
Wires (especially inaccessible wiring bundles and wiring behind the control panel)			
Circuit breakers (check for proper operation at rating)			
Capacity check of aircraft battery			
EMPENNAGE			
Condition of stabilizer attachment structures			
Inaccessible areas of stabilizers			
ENGINE			
Accuracy of engine indicating gauges			
Engine mounts			
Internal condition of exhaust piping			
Condition of exhaust clamps and hardware			
Turbocharge wastegate controller bench			
Accuracy of turbocharger system indicators			
FUEL SYSTEM			
Plumbing			
Fuel quantity			
Fuel tank installation			
FUSELAGE			
Seat tracks			
Fuselage attach points			
Inaccessible areas			
INSTRUMENTS			
Plumbing			
Instrument panel shock mounts			
Pitot/static system			
Airspeed/vertical airspeed calibration			
Altimeter calibration			

Table 4-4-1. Sample of a general aviation aging aircraft inspection checklist

COMPONENT	SERVICE LETTERS AND/OR ADVISORY MATERIAL	DATE LAST INSPECTED/REPLACED	FINDINGS/ NOTES
Stall warning system calibration			
Magnetic compass calibration			
LANDING GEAR			
Attach bolts NDT			
Axle/attach NDT			
Inaccessible areas of wheel well			
MODIFICATIONS			
Records review for proper approval of all modifications including weight and balance revisions			
Additional continued airworthiness inspection/maintenance requirements for modifications that do not include instructions for continued airworthiness			
REPAIRS			
Records review for proper approval of all repairs including weight and balance revisions			
SYSTEMS			
Cabin pressure control accuracy			
Heater muff type heating system			
Seat belts (certify or replace)			
Cargo restraints (certify or replace)			
Fire extinguisher (certify)			
Hydraulic system plumbing in inaccessible areas			
Hydraulic system gauging accuracy			
Proper operation of oxygen regulator			
Hydrostatic check of oxygen bottle			
Calibration check of oxygen auto dispensing system			
Pneumatic plumbing			
Pneumatic gauging accuracy			
Calibration check of pneumatic de-ice system pressure indication (switches or gauges)			
Vacuum plumbing			
Vacuum regulation			
Vacuum gauging accuracy			
WING			
Wing attach bolts			
Strut attach fittings			
Front & rear strut			
Strut attach bolts			
Spar			
Nails (wood spar)			
Compression struts			
Drag wire assembly			
Ribs/wingtip			
Fabric			
Inspection holes			
Drains			
Wing general			

Table 4-4-1 (cont'd). Sample of a general aviation aging aircraft inspection checklist

Wood and composite structures don't sustain fatigue damage the same as metal does, but repeated loads still have long-term damaging effects.

Good records research enables an assessment of the individual airplane history including particular and pertinent environmental and usage factors. When coupled with research results for the model type history, an individual airplane assessment becomes more meaningful.

Special inspections. An assessment of an airplane's paperwork is only a prelude to a thorough aging evaluation. For aging airplanes, the normal annual inspection minimum requirements specified in 14 CFR 43.15, Appendix D, or those recommended by the manufacturer are probably not enough. A detailed inspection or series of inspections, modifications, parts replacements, or a combination of these may be necessary to keep an aging airplane operating safely.

As an airplane ages, the inspection methods and techniques may change from what was previously required. Special inspections may be required because of high aircraft time, severe operation and inactivity, outside storage, modifications, or poor maintenance. The records research will provide information needed for owners and mechanics to determine what may be needed for a particular airplane or airplane type.

Table 4-4-1 is an example of an inspections checklist for general aviation aircraft.

Chapter 5
AVIONICS corrosion

Left: Avionics inspections require detailed knowledge of more than just radios.

Section 1
Preventive Maintenance Programs

Successful avionics cleaning and corrosion prevention/control efforts depend on a coordinated, comprehensive preventive maintenance program. Everyone involved with the operation, repair and maintenance of avionics equipment must take an "all hands" approach to cleaning, inspection and corrosion prevention and control.

Recognizing the appearance of corrosion or corrosion products for specific metals is an important part of an avionics corrosion prevention and control program. Metals are susceptible to corrosion because all metals have a tendency to return to their natural forms. For example, iron tends to return to iron oxide (rust). Avionics systems make use of many metals not normally considered for airframe structures. Some of the rarer metals are found in avionics components, in transistors, miniature and micro-miniature circuits and integrated circuits. Table 5-1-1 lists the metals most often used in electronics and avionics components. In addition to recognizing the appearance of corrosion or corrosion products for specific metals, maintenance personnel must be able to recognize a corrosive attack from solder flux, microbes, insects and/or animal attack.

Corrosion Effects on Metals

Deterioration (corrosion) of a metal is caused by a chemical reaction with its environment. The corrosive effect can be accelerated by many factors. No metal can have a perfect environmental integrity and therefore will corrode. Corrosion of a specific metal will take on many different

Learning Objectives:
- Cleaning and Preservation
- Paint Damage
- Sealant Applications
- Antennas
- Electromagnetic Interference

forms as the corrosive attack progresses. The following is a short description of corrosion with respect to the most commonly used metals in avionics systems. Chapter 1, Elements of Corrosion covers in detail the types and effects of corrosion on metals.

Iron and steel. Iron and steel are used in avionics components as leads, magnetic shields, transformer cores, brackets, racks and general hardware. Some of these components are plated with nickel, tin, or cadmium. Iron and steel surfaces are normally protected by applying a plating system, paint system or application of a preservative compound.

Corrosion resistant steel. Stainless steel is used for mountings, racks, brackets and hardware in avionics systems. Stainless steel is not readily susceptible to corrosion because of a tough chromium oxide film on the surface. Stainless steel is susceptible to crevice corrosion and exposure to salt water will cause pitting of the surface.

Aluminum alloys. Aluminum and aluminum alloys are widely used in avionics systems for electrical connectors and back shells, cabinets, housings, chassis, structures and mounting fixtures. In most environments, especially moist salt-laden air, aluminum alloys are subject to many types of corrosive attack and therefore require protection. Aluminum surfaces are protected by an entire paint coating system composed of a chemical conversion coat, primer and topcoat.

Magnesium. Magnesium alloys are used throughout avionics systems because of their light weight. Magnesium alloys can be found in antennas, component structures, chassis, supports and frames. Magnesium is highly susceptible to corrosion when exposed to any environment without a protective coating. Magnesium forms a strong (anodic) galvanic cell with every other metal and is always the metal that will suffer the corrosive attack. When corrosion is found on magnesium, prompt corrective action is required.

Copper. Copper and copper based alloys are used extensively in avionics systems in wiring, contacts, springs, connectors and printed circuit boards. Copper and copper based alloys (brass and bronze) are quite resistant to corrosion, since they are cathodic to most of the other metals.

Cadmium. Cadmium is used primarily in avionics equipment as a plating on hardware (nuts, bolts, etc.) and electrical connectors. It is also used to provide a compatible surface for parts in contact with other material. Cadmium, when plated over steel, is anodic to the steel and protects the steel as a sacrificial coating.

Silver. Silver is normally used as plating material over copper in wave guides, miniature and microminiature circuits, wiring and contacts. It is also used on radio frequency (RF) shielding. Silver does not corrode in a normal sense, but it will tarnish in the presence of sulfur. The tarnish (silver sulfide) appears as a brown to black film. Corrosion treatment should be limited to cleaning.

When silver plating over copper is damaged, there can be an accelerated corrosive attack of the underlying copper. The "red plague" is readily identifiable by the presence of a brown-red powder deposit on the exposed copper. In the case of wiring, the problem is compounded by the wire insulating material. The insulating material can prohibit detection of the damaged silver plating until the damage is extensive.

Gold. Gold is the best plating material for electrical connections because of its corrosion resistance and the ease with which it can be soldered. Gold is used on printed circuits, semiconductor leads and contacts and is usually plated over nickel, silver, or copper. Gold is a noble metal and does not normally corrode; however, a slight tarnish will appear as a darkening of its normally reflective surface. When gold is plated over silver or copper, accelerated corrosion can occur at pin holes or pores in the gold plating. This tarnishing of the gold over silver is readily identified by a brown to black film. The tarnishing of the gold over copper is identified as a blue-green film. The methods employed in tarnish removal are critical on gold-plated components because the plating is very thin (typically 0.00015 inch thickness).

CORROSION-RESISTANT METALS	NON CORROSION-RESISTANT METALS
Brass	Aluminum
Bronze	Antimony
Copper	Arsenic
Gold	Beryllium
Iridium	Bismuth
Monel	Cadmium
Nickel	Cobalt
Palladium	Germanium
Rhodium	Indium
Silver	Iron
Stainless Steel (CRES)	Lead
Tantalum	Lead-Tin Alloy
Tin	Magnesium
	Mercury
	Steel
	Tungsten

Table 5-1-1. Metals most commonly used in avionics systems

ORGANIC MATERIALS	OTHER MATERIALS
Asbestos	Acrylics
Cloth	Adhesives
Cork	Conformal Coatings
Felt	Elastomers
Glass	Encapsulates
Leather	Laminates
Paper	Lubricants
COMPOSITE MATERIALS	Paint
Ceramics	Plastics
Graphite	Potting Compounds
Polymers	Primers
	RTV
	Sealants
	Silicone
	Tapes

Table 5-1-2. Nonmetals most commonly used in avionics systems

Purple plague is a brittle gold-aluminum corrosion product formed when a gold-plated component and an aluminum component are mechanically attached or bonded together. Microelectronic circuit failures can occur at the interconnecting mechanical bond as this gold-aluminum corrosion product grows.

Tin. The use of tin in solder is a well-known application. Tin is also used as a plating on RF shields, filters, crystal covers and automatic switching devices. Tin is the best solder and corrosion-resistant coating of the available metallic coatings. The problem with tin is its tendency to grow "whiskers" on tin-plated wire and other tin-plated devices, to an extent that they will cause shorting across microelectronic circuits.

Black plague is a black substance that forms in the liquid cooling systems of high power radars. This substance adheres to the walls of tubing and components in the cooling system and affects the heat transfer characteristics of the system.

Nickel. Nickel is primarily used as an electrolysis coating and is subject to pitting corrosion. Flaking of the nickel coating can also occur when the underlying metal corrodes.

Corrosion on Nonmetals

Deterioration of nonmetallic subassemblies and other hardware costs commercial and private operators millions of dollars per year in replacement material costs and loss of equipment availability. In most cases, the deterioration of the nonmetallic material permits the intrusion of moisture into the equipment. This deterioration creates physical swelling, distortion, mechanical failure through cracking, altering of electrical characteristics, etc. The most common nonmetals used in avionics systems and the nature and appearance of their deterioration are listed in Table 5-1-2.

Corrosion effects of solder flux. Solder flux residues may be conductive and corrosive. They are often "tacky"; collecting dust which can absorb moisture and create current leakage paths. Solder flux resin appears as an amber-colored globule, drip, or tail at or near the solder joint. Under ultraviolet light, traces of flux appear as a fluorescent yellow to light brown residue. When soldering, use the lowest flux content possible (even "neutral" fluxes have some acid in order to remove metal oxides). After soldering, all flux residue must be completely removed by cleaning. Complete removal can be verified by the use of an ultraviolet light.

Effects of organic materials. Bacteria and fungi not only feed on organic material, but also release acids which are corrosive. Bacteria and fungi may be found on encapsulates, conformal coated circuit boards, rubber gaskets, thermoplastics, optical lenses, etc. The presence of bacteria and fungi can be readily identified by damp, slimy and bad-smelling growths. These growths vary in color from black, blue-green and green to yellow.

Effects of insect and animal attack. Small insects and animals may enter parked aircraft and packaged equipment. They often feed on organic materials such as polyethylene and wire insulation that result in system or equipment failures. The presence of nests, holes and excrement indicate animal or insect attack as shown in Figure 5-1-1. This problem is generally more severe in equipment that has been in storage or has been out of service for a long period of time.

Dust and lint. Avionics equipment is subject to dust and lint accumulation. This condition is generally evident when the equipment has been installed for long periods of time and can become more severe with forced cooling air. In addition to accumulation from the movement of air, dust and lint can be attracted by magnetic fields from electric currents surrounding wiring and equipment. Dust can also accumulate on the top and interior surfaces of components, shown in Figure 5-1-2.

The problem with the accumulation of dust and lint is that it will trap and hold moisture which can provide an electrolyte for corrosion and fungus growth. Additionally, dust and lint can degrade avionics equipment by being conductors or insulators. When dust and lint act as conductors in the presence of moisture, they provide a path for a current flow to either ground as a short circuit or an unwanted circuit path between components. When dust and lint act as insulators, avionics equipment can overheat, causing premature failure. Frequent inspections and general cleaning of equipment will control the accumulation of dust and lint.

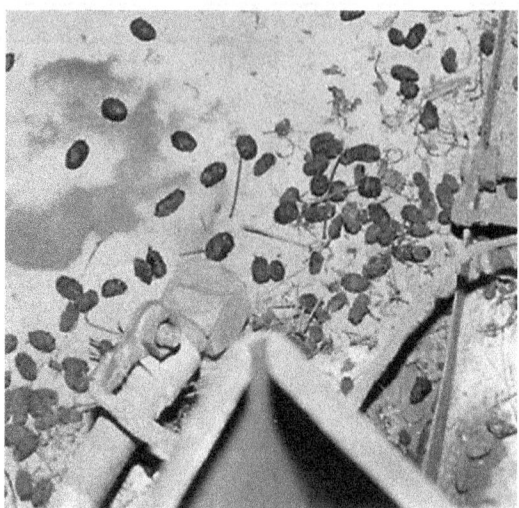

Figure 5-1-1. Rodent droppings in a bilge area

Section 2

Cleaning and Preservation

The materials, equipment and techniques described in this section are intended to assist the technician at a maintenance or avionics repair facility. This section discusses cleaning and minor repair requirements of avionics equipment. A specialized facility is often needed to perform extensive cleaning and major repairs of avionics.

> **NOTE:** *Many of the components in avionics equipment require specialized training and certifications. Before any piece of equipment is opened the proper reference materials, tools and training must be available.*

Required Facilities

At a minimum, an avionics cleaning and repair facility should include the following resources for the cleaning, drying, preserving, packaging, handling and shipping avionics equipment:

- Adequate lighting and a temperature/humidity controlled ventilation system
- Adequate space for safe operation of avionics cleaning and corrosion control equipment
- Operating instructions for each piece of equipment
- All Material Safety Data Sheets (MSDS) for materials used
- Safety equipment and protective personal equipment as required by local, state and federal ordinances
- Personnel trained in the recognition of corrosion on avionics equipment
- Personnel trained in the safe and proper operation of specialized equipment.
- Quality assurance inspectors trained in the operational characteristics and restrictions of each piece of equipment

Cleaning Compounds

Avionics technicians must understand the functions, capabilities and restrictions associated with each piece of specialized support equipment and the hazards of the cleaning and corrosion removal products and preservation materials. This knowledge will prevent injury to personnel and damage to avionics equipment and support equipment.

A list of the various cleaning compounds and solutions, their characteristics, application, mixing instructions and restrictions follows:

> **CAUTION:** *Some materials can create hazardous conditions or damage equipment if not used strictly in the applications and manner described. Some of the materials are chlorofluorocarbons (CFC5), an ozone depleting substance (ODS). These materials may be banned from use by environmental restrictions.*

Aircraft cleaning compound MIL –C-85570. This is a general purpose cleaner for light soil and dirt. It can be used in equipment bays, external covers and antennas. This cleaner must not be used around oxygen fittings or regulators due to the possibility of explosion or fire.

Liquid detergent MIL-D-16791. Liquid detergent can be used to clean transparent covers and glass. It is also used in a solvent spray booth and aqueous ultrasonic cleaners.

Electrical contact cleaner MIL-C-83360. A spray cleaner and lubricant that is compatible with potting compounds, rubber gaskets and insulation. May be used as a lubricant for electrical contacts. It should not be used in areas that require soldering or coating.

Dry cleaning solvent PD-680. A general purpose cleaner for heavy dirt, dust and contaminants. It is used in equipment bays, external cases, equipment racks and mounts. It is applied by wiping or brushing. It is not to be used around oxygen fittings or regulators.

Isopropyl alcohol. This is used as a general purpose cleaner to remove salt residue, solder flux and light oil.

Distilled or fresh water is used to dilute isopropyl alcohol or detergents for use in cleaning and as a final rinse. Water may also be used to remove dust, dirt, salt deposits and cleaning solutions.

NOTE: *Many of the materials identified are hazardous and toxic to personnel not using appropriate personal protective equipment. These materials can also be potentially damaging to avionics equipment and aircraft if used in improper concentrations or misapplied. Prior to using any chemicals such as paint strippers, detergents, solvents, conversion coatings, primers, or paints, personnel should review the appropriate Material Safety Data Sheets (MSDS) for warnings and cautions or the hazardous material identification labels on the containers.*

Avionics Cleaning Equipment

The various types of avionics cleaning equipment, their uses and specific restrictions are described in the following paragraphs:

NOTE: *Technicians should refer to the appropriate equipment service manual for specific operating instructions. Ultrasonic cleaning of printed circuit boards (PCB) is generally not authorized due to the difficulty in determining which component on the board will be susceptible to damage.*

CAUTION: *Miniature and microminiature PCBs may be susceptible to damage from the ultrasonic frequency, power level, or both. Sealed bearings, synchro and servo bearings, instrument bearings and similar devices with permanently lubricated (sealed) bearings can experience cleaning solution intrusion and the removal of the lubricant. These components could be rendered useless unless the processing procedures protect those components from degeneration or the lubricant is replaced.*

Sealed components (other than hermetically sealed) can trap the detergent and water solution. This may cause drying problems. In each case, the sealed component should be opened and inspected for trapped water.

Cleaning booth. Water base solvent spray booth is used for the removal of dirt, dust, salt spray deposits and loose corrosion deposits. The cleaning action is accomplished by a detergent and spray system.

Spray equipment provides an air pressure powered spray of a detergent/water solution through a hand held spray gun. Water can be either filtered or tap water. The equipment can also deliver rinse water or a drying jet of air.

Some spray booths include a turntable which allows 360° rotation of the avionics equipment being cleaned or rinsed.

The preferred detergent for the water base solvent spray booth should conform to manufacturer's

Figure 5-1-2. Accumulated dirt and dust can act as insulators, raising the heat level of the equipment.

specifications. It should be mixed in accordance with the manufacturer's instructions.

Aqueous ultrasonic cleaner. This type of cleaner is used for the removal of dirt, dust, salt spray deposits and loose corrosion deposits. The cleaning action is accomplished by the ultrasonic scrubbing action of the detergent and water solution.

- The maximum operating temperature should not exceed 130°F (54°C).
- The maximum operating frequency used should be 20 kHz.
- Only use the recommended liquids in the ultrasonic cleaner.

Paper capacitors and paper bound components disintegrate in the detergent and water solution. Thin metal foil types of gummed labels can loosen and separate in an ultrasonic cleaner.

Solvent ultrasonic cleaner. The solvent ultrasonic cleaner is used to remove light to heavy oil, grease and hydraulic fluid contamination by ultrasonic scrubbing in a solvent solution. Use only an approved solvent in the solvent ultrasonic cleaner. The maximum operating temperature should be the solvent's boiling point. The maximum operating frequency should be 40 kHz.

Solvent degreasing, solvent vapor rinsing and solvent vapor drying are additional functions of the solvent ultrasonic cleaner. Solvent degreasing is performed by solvent ultrasonic action in the degreaser tank. Solvent vapor rinsing (part of the degreasing function) is performed by a solvent vapor cloud. The solvent vapor cloud is created by a cooling coil placed near the top of the degreaser tank. Solvent vapor drying is also performed by the solvent vapor cloud.

Some solvent ultrasonic cleaners have a solvent degreaser tank. The solvent vapor rinse and solvent vapor drying function do not use the ultrasonic frequency function. Therefore, they may be used to rinse and dry PCBs.

> **NOTE:** *Some acrylics may be susceptible to damage from the solvents used in solvent ultrasonic cleaners. Coaxial connector gaskets and other neoprene rubber components are susceptible to damage by some solvents. Refer to the manufacturer's instructions or use another non-solvent cleaning process.*

Abrasive cleaning tools. Abrasive cleaning tools such as portable mini-abrasive units and blast cleaning cabinets are used to remove corrosion and corrosive products.

The abrasive blast medium used in abrasive cleaning tools can easily be trapped in miniature and micro miniature female edge connectors. When working on components where these connectors are installed, the connectors should be sealed with pressure sensitive tape.

Delicate metal surfaces such as metal plating are susceptible to damage from indiscriminate use of abrasive cleaning tools. Only properly trained personnel are authorized to use abrasive cleaning tools on avionics equipment.

Some miniature and micro miniature PCBs contain electrostatic discharge sensitive (ESD) devices. These ESD sensitive devices may be destroyed by the static charge created in the rapid movement of air and abrasive agents in the abrasive cleaning tools. For this reason, abrasive cleaning tools should not be used where ESD sensitive devices are installed.

Avionics Cleaning Procedures

Corrosion products and contamination from other sources, described in Chapter 1, Elements of Corrosion, as well as in Section 1 of this chapter, are responsible for numerous problems and failures in avionics equipment. Proper cleaning can prevent many of these problems and is the next logical step after an initial inspection. Cleanliness is very important in maintaining the functional integrity and reliability of the avionics system and components. Corrosion products and contaminants may be either conductive or insulating. As conductors they may provide undesirable electrical paths and as insulators they may interfere with the avionics equipment or systems operation.

Cleaning method selection criteria. The best technique for selecting a cleaning method is to select the mildest cleaning method that will accomplish the task. The selection of the cleaning method is a decision that may be outlined in an original equipment manufacturer's (OEM) maintenance manual or made by authorized personnel at the avionics equipment repair facility. More extensive cleaning and removal methods are covered in Chapter 3, Corrosion Control. The decision on the mildest cleaning method should consider the following:

- Certain circuit components can be damaged by support equipment and cleaning solutions and solvents
- Type and extent of the corrosion damage or contamination
- Accessibility to the corrosion damage or contamination
- Type of avionics equipment

Hazards of cleaning. As previously mentioned, it is a good maintenance practice to use the mildest cleaning method that will ensure removal of the corrosion or contamination. It is also important that the correct cleaning solutions and cleaning materials be used to avoid further damaging the avionics equipment.

Cleaning solvents or other cleaning materials can become trapped in crevices or seams of components or chassis. These trapped cleaning materials could then interfere with the application of protective coatings and result in the initiation of corrosion.

Vigorous or prolonged scrubbing of laminated circuit boards can cause damage to the boards. Certain cleaning solvents can soften conformal coatings, wire insulation, acrylic panels and PCBs.

When to clean. The immediate removal of corrosion or contaminants on avionics equipment and the surrounding structure is a high priority in a proper corrosion control program. Immediate cleaning should be accomplished on avionics equipment and components if they have been exposed to adverse weather, salt water immersion or spray, fire extinguishing agents, spilled battery electrolytes, other acids, high pH alkaline cleaners, mercury and corrosion products during component repair.

Pre-cleaning criteria and cautions. Some important precautions which should be taken before cleaning are:

1. Disconnect electrical and other power sources (mechanical/hydraulic).
2. Ensure the work area and equipment are safe for maintenance.
3. Ensure drain holes are open.
4. Remove covers and panels for accessibility.
5. Disassemble where practicable.
6. Use only authorized and compatible materials.
7. Mask and protect accessories or components to prevent intrusion of water, solvents and cleaning solutions.

Cleaning and drying restrictions. Certain circuit components create potential problems during cleaning and drying. These problems can generally be overcome prior to cleaning by carefully masking those components likely to trap and hold cleaning liquids due to their construction. Mechanical, shock and heat damage are other types of damage that can occur during cleaning and drying. To avoid problems, seal small components with pressure sensitive tape.

Seal large components in static-free plastic bags or other water- and vapor-proof barrier material. Place the plastic bag or barrier material around the component and seal with pressure-sensitive tape. When possible and if authorized, remove subassembly components and clean them separately.

Hand cleaning methods. Hand methods should be used for cleaning small, delicate, confined surfaces where parts cannot tolerate other cleaning methods. Also, hand cleaning methods should be used when accessories/facilities for other methods are not available. Some of the equipment that may be utilized for hand cleaning includes lint-free cloth, cheesecloth and cotton tip applicators. More vigorous cleaning can be accomplished with acid brushes, toothbrushes or other soft bristle brushes. A plastic manual spray bottle is very useful for spot spraying of cleaning solvents and for rinsing.

Fingerprint removal. The salts and oils from human fingerprints are highly corrosive. Apply a mixture of isopropyl alcohol and distilled water to the affected area with a clean cloth, cheesecloth, acid brush, or toothbrush as appropriate.

> **CAUTION:** *Do not use synthetic fiber wiping cloths with isopropyl alcohol due to its low flash point. Dry fiber wiping cloths will cause a static charge to build up and may result in a fire.*

1. Wipe or scrub affected area until contaminants have been removed.
2. Remove residue by wiping or blotting with clean cheesecloth or a cotton tip applicator as required.
3. Inspect area for signs of residue or contaminants.
4. Repeat the process as required until all evidence of contamination is removed.

Cleaning and removal of solder flux residue. Solder flux residues should be removed from circuit boards and circuit components using a solution of distilled water and isopropyl alcohol to clean the affected area. After a soldering operation, clean the affected area and one inch around the solder point to ensure complete decontamination.

> **CAUTION:** *Lead contained in solder can rub off onto a person's fingers from a soldered joint. Lead and lead oxide are toxic and cannot be eliminated from the body. This toxic poison will accumulate in the body. Touching solder followed by smoking or eating is a potential means of ingesting trace amounts of lead. Wash hands thoroughly following any soldering or desoldering operation.*

Visually inspect circuit boards and circuit components for evidence of soldering flux residue using an ultraviolet light source. Repeat the

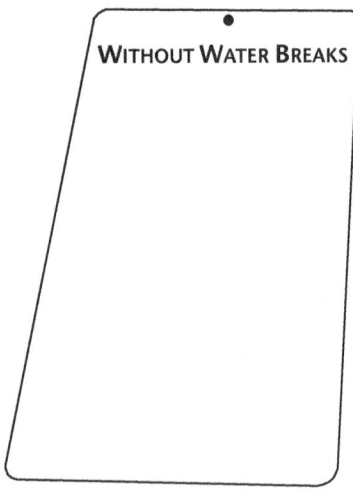 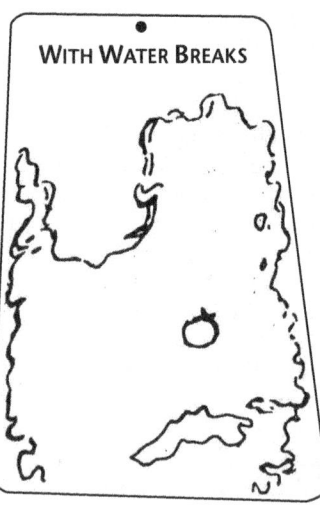

Figure 5-2-1. A water-break free surface compared with one with breaks

cleaning process as required until all evidence of solder flux is removed.

> **CAUTION:** *Prior to using the ultraviolet light for inspection to ensure complete removal of soldering flux residue, examine the piece of equipment for Erasable Programmable Read Only Memory (EPROM) components. EPROM have windows that are usually covered with an aluminum foil mask or black tape. Visually examine to ensure the tape or foil has not lifted from the window. Ultraviolet light is very intense and can degrade EPROM devices.*

Surface Preparation

It is critical that bare metal surfaces be completely clean to ensure adhesion of any subsequent coating material. Examples of subsequent coatings include: chemical conversion coating (pre-paint), paint system, sealant, dry film lubrication, etc. The water-break test is commonly used to identify a surface that is clean enough for coatings to adhere.

- Prepare a surface by feathering the edges of any existing paint finish to ensure a smooth overlapping transition between the old and new paint coating and rinsing with fresh water. Particular attention should be given to fasteners and other areas where residues may become entrapped.

- Visually inspect the part after the rinse; the surface should be water-break free. A surface showing water breaks (water beading or incomplete wetting) is contaminated, usually with grease or oil (fingerprints, etc.). The contaminated surface will not allow proper adhesion with any subsequent protective coating system (conversion coating, primer, topcoat). Figure 5-2-1 illustrates a water-break surface.

- If the surface is not water-break free, clean the area again by lightly scrubbing the area with very fine or fine abrasive mat saturated with cleaning solution.

- Rinse the surface with water.

- Visually inspect the part after the last rinse. The surface should be water-break free.

Post-cleaning procedures. After completion of the cleaning steps, inspect the affected area for signs of residue, surface film, or water. If inspection reveals the area is not clean, (residue, surface film), repeat the cleaning procedures. Water-displacement, preservation and lubrication should follow the cleaning and drying steps for completing preventive maintenance.

Drying Equipment and Procedures

Drying time depends on the complexity of the equipment/component being dried. The more complex the individual component, the longer the drying time. Another consideration in drying time is the humidity or moisture content of the air where the drying oven is operated. The higher the moisture content of the ambient air, the longer the drying time.

> **CAUTION:** *Portable air blowers, hot air blowers, hair dryers and similar drying devices may cause fires when used in or around aircraft. Use only authorized (spark proof) hot air guns. The motion of air from an aerosol spray, compressed air and air from dryers can generate static charges that can degrade or destroy Electrostatic Discharge (ESD) sensitive devices. Care must be exercised during handling, cleaning and repair of these items.*

Drying preparation. Prior to placing a component in a drying oven, remove all covers, lids, etc. Ensure that any pressure-sensitive tape and protective plastic bags used during the cleaning cycle have been removed. Air drying is usually adequate for housings, covers and some hardware. This method is not considered adequate for more complex equipment or components that may contain cavities or moisture traps.

Drying with hot air blower. Blow off excess water with dry air or dry nitrogen at not more than 10 p.s.i. pressure. Deflect air off interior back and sides of enclosure to diffuse air jet. Dry the equipment with a hot air gun similar to the type shown in Figure 5-2-2.

Drying with circulating or forced air oven. The circulating air drying oven is used to dry small electrical and electronic components, such as non-pressurized instruments, control boxes, PCBs and similar devices. The circulating air drying oven should not be operated above 130°F (54°C) when drying avionics equipment or components. Damage may result from overheating of discrete electronic circuits components. After placing the component in the oven, set the thermostat and close the door.

If a timer is available, set the time for approximately 3 to 4 hours and upon completion of the drying cycle, remove the component(s).

Solvent vapor drying. Solvent vapor drying is an additional feature of the solvent ultrasonic cleaner. This method of drying is considered the fastest and most efficient method of drying avionics components. The drying time is usually 15 seconds to 3 minutes. Follow the equipment manufacturer's operating procedures for cleaning and degreasing.

> **CAUTION:** *Environmental regulations may prohibit the use of the ozone depleting chlorofluorocarbons (CFC) solvents.*

1. Suspend the component being processed in the solvent vapor cloud for a period of 15 seconds to 3 minutes. Drying time depends on the complexity of the component and the amount of water or solvent present.
2. Rinse by altering the position of the component in the vapor cloud to drain any trapped water or solvent.
3. After thorough rinsing, slowly withdraw the component out of the vapor cloud.
4. Inspect the component for visible signs of water or solvent.
5. If required, repeat the solvent vapor drying process.

Preservation and Lubrication

Surfaces and components not normally conformal coated or painted need preservation. Cleanliness and elimination of moisture are keys to avoiding corrosion. Preservation of equipment is essential since it is nearly impossible to guarantee a dry, moisture-free environment. In today's avionics systems, miniaturization has resulted in very small electronic circuits. Even a small amount of corrosion can cause the entire system to fail.

Preservation has become an essential part of the repair and maintenance of avionics systems. Preservation is done to protect nonmoving parts by filling air spaces, displacing water and to provide a protective coating. Preservation also protects components such as hinges, control cables, gears, linkages, bearings, etc. from wear by providing lubrication.

What to preserve. Preservatives should be used only where their application and maintenance will not hamper electrical circuits or component operation. Most preservatives form a nonconductive film that acts to insulate mating surfaces from moisture intrusion. The following avionics equipment and components may require preservation and maintenance:

- Door latches and hinges
- Electrical connectors (internal and external)
- Shock mounts, rigid mounts and associated attaching hardware and brackets
- Any dissimilar metals not protected by other coating systems
- Antenna mounts, brackets, hardware and housings
- Fasteners, screws, nuts and bolts
- Equipment lids on interior or exterior of equipment that is susceptible to moisture
- Solder joints not otherwise protected by other coatings
- Any unprotected surface that will not receive a paint coating system or other coating material (plating or conformal coatings)
- External and internal surfaces of coaxial connectors
- External surfaces of cooling system joints
- Ground straps and wires

What not to preserve. The following items should not be preserved or come in contact with preservatives:

- Laminated circuit boards that are conformal coated

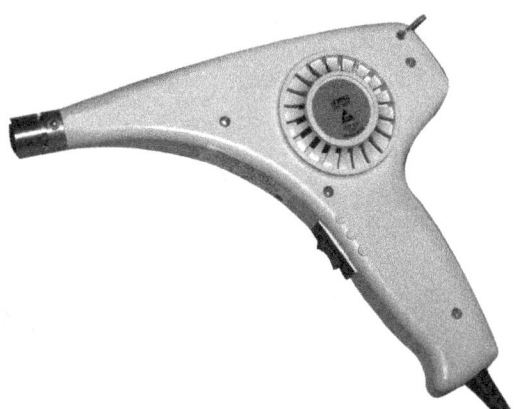

Figure 5-2-2. When drying equipment with an air gun, surfaces should not reach temperatures above 130°F (54°C).

Figure 5-2-3. Shipping containers such as the one above protect components from vibration and shock

- Nonmetallic surfaces such as acrylic control box face plates
- Tunable capacitors and inductors
- Internal surfaces of wave guides
- Internal surfaces of tuned tanks
- Relay and circuit breaker contacts
- Fuses

Preservative materials. Preservatives are materials that can take the place of more permanent coating materials such as paint, but require removal and repeated application on a scheduled basis. Some preservatives may also act as water-displacing compounds. Preservatives must be applied over water displacing, corrosion-preventive compounds to accomplish a complete water displacing and preservative system on all areas exposed to elements and moisture.

Lubricants. Lubrication of equipment performs several important functions: it prevents wear between moving parts, provides a barrier to corrosion and is a short-term preservative. Particular attention should be given to lubrication points around hinges, latches, etc., for prevention of lubrication breakdown. Lubrication breakdown includes caking of the lubricant, an indicator of contamination.

Requirements. Maintenance personnel should refer to the applicable maintenance manual for specific lubrication requirements, materials and frequency.

Packaging, Handling and Storage

An avionics corrosion control program must include procedures for packaging, handling and storage of avionics equipment and components. These components will be rendered useless if the packaging, handling and storage procedures are not followed. The materials used must be compatible with the avionics equipment and the anticipated environment. Figures 5-2-3 and 5-2-4 show typical storage and shipping containers.

Packaging and storage materials guidelines. Certain packaging materials containing wood, cotton, foam and paper are susceptible to mold and fungal attack. These materials and other items such as shredded newspaper, excelsior and cardboard may give off sulfurous or acidic vapors that can promote a corrosive attack on electronic components.

- Use only metal or preserved wooden shelves for storing avionics equipment and components.
- Provide closed-cell polyethylene foam 1/2 inch thick as a cushioning for equipment on shelves, pallets, etc. Do not use an open-cell foam or sponge rubber or similar material that will hold moisture.
- Electrical connectors should have conductive plastic or metal caps for electrical protection.
- Use cellular plastic film cushioning material (bubble wrap) for short-term protection during transportation of equipment and to protect against handling and shock damage.

NOTE: *Never place bubble wrap in direct contact with ESD-sensitive devices. The electronic device or component should first be placed in a conductive bag.*

Use plastic bags for short-term protection of non-installed small components and micro miniature PCBs against moisture and contamination.

Handling. Damage can occur to avionics equipment because of incorrect packaging methods and rough handling between the manufacturer, aircraft and avionics repair facilities. The best method of avoiding handling damage when trans-

Figure 5-2-4. Bubble wrap should be used for short-term storage

porting equipment is to use specially designed cushioned shipping container. A method that can be used if the original shipping container is not available is to use cellular plastic film cushioning material (bubble wrap). Bubble wrap is primarily used to absorb shock and is not intended as a water vapor-proof packaging material.

Electrical connector and wave guide caps. Metal caps and blank off plates, preferably moisture and vapor proof, are the preferred methods of protecting electrical connectors and wave guides from contamination and damage. Conductive plastic caps are another acceptable method of protection. Use only plastic caps which cover by surrounding the connectors or wave guide. Do not use push-in type caps or covers. These type caps or covers can easily be overlooked during assembly or become foreign object debris (FOD). When metal or conductive plastic caps and blank off plates are not available, electrical connectors and wave guides may be capped off with pressure-sensitive tape.

Desiccants. Desiccants are normally packaged with equipment packaged for storage or shipment. The purpose of the desiccant is to absorb moisture and lower the relative humidity. Typical desiccant bags are shown in Figure 5-2-5.

> **CAUTION:** *Do not use loose desiccants in packing of avionics equipment. Loose desiccant may contaminate and cause damage to the packaged equipment.*

The following considerations apply to desiccants:

- Desiccant material should be contained in rupture-resistant sturdy bags.
- Desiccant bags should be secured to prevent movement.
- Desiccant bags should not be placed on or permitted to come in contact with unprotected surfaces
- Desiccants should be reactivated prior to reuse.
- Desiccant bags should not be removed from their sealed container until ready for use.

If a desiccant bag should break open during transit, clean the avionics equipment immediately upon discovery. Do not turn moving parts any more than absolutely necessary until all desiccant particles have been removed. Work out the desiccant particles with a clean acid brush and not more than 10 p.s.i. dry air pressure.

Humidity indicators. Humidity indicators should be placed in containers when desiccants are used. A humidity indicator is used to determine if a desiccant is sufficiently active to maintain an acceptable relative humidity within the container.

> **CAUTION:** *Do not place humidity indicator in direct contact with metal. Chemicals used in the indicator may cause corrosion.*

Section 3

Corrosion Removal, Painting and Sealing

This section outlines the materials, equipment and techniques involved in corrosion control for avionics equipment. More extensive corrosion is removed using the techniques discussed in Chapter 3, Corrosion Control. Maintenance personnel should analyze each corrosion problem and select the correct corrosion removal and preservation materials. Where possible, follow-up actions should be conducted to ensure that all corrosion has been removed and proper protection has been applied. An avionics corrosion control program is an important function in maintaining any aircraft and aircraft component.

The program requires knowledge of the science and technology of avionics corrosion control. Preventive maintenance must occur as part of all maintenance functions performed on avionics systems. Whenever equipment is removed from the aircraft for bench check or repair, covers and housings should be inspected and treated for corrosion. Avionics technicians must ensure that corrosion, repair, treatment and preventive maintenance become a normal part of their maintenance and repair procedures.

Figure 5-2-5. Silica gel, or dessicant, keeps moisture levels low during storage and shipping.

ITEM	SPECIFICATION	USE
Polysulfide synthetic rubber, accelerated cure sealing compound	MIL-S-8516	Low voltage electrical connectors, wiring, where temperature is below 200°F (93°C). Not for use in engine bays, keel areas, or hot bleed air ducts.
Silicone rubber, RTV (Room temperature vulcanizing)	MIL-S-23586 Type II, Class 2 Grade A	Small electrical connectors not in closed spaces, where maximum temperature exceeds 450°F (232°C).
Silicone rubber	MIL-S-23586 Type II, Class 2, Grade B	Electrical connectors in closed spaces where temperatures exceed 200°F (93°C). Maximum temperature of 450°F (232°C).
Silicone adhesive, RTV	MIL-A-46146 Type I	General purpose one part adhesive sealant used for encapsulating and sealing electrical and electronic components. Good resistance to oxidation, weathering and water.
Silicone rubber sealant, RTV	MIL-A-46146 Type II	Less viscous one part compound for sealing electrical connectors.
Silicone adhesive, sealant, RTV	MIL-A-46146 Type III	Non-corrosive, one part sealant used to encapsulate and seal electrical and electronic components between 250 and 350°F.
Two-part polysulfide, fuel resistant sealing compound	MIL-S-8802	Used in areas where fuel and oil is present to prevent entry of corrosive elements. Maximum temperature 275°F (135°C).
Polysulfide sealing compound	MIL-S-8802, Type A	Used for sealing gaps, seams and faying surfaces. Excellent resistance to water, aircraft fuels and petroleum based oils. Good resistance to ester base hydraulic fluids. Adheres to most metals and aircraft materials. For temperatures up to 250°F (121°C).
Two part polysulfide with soluble chromates, corrosion inhibiting sealing compound	MIL-S-81733	Used to prevent the entry of corrosion elements into faying surfaces subject to galvanic action. Maximum temperature of 250°F (121°C) Resistant to shrinkage, ester type hydraulic fluids. Four types allow long assembly times.
Two-part polyether polyurethane sealing compound	MIL-M-24041	Used for seating and reinforcing electrical connectors, in a seawater environment. Will adhere to metal, rubber and PVC. Temperature range from -80 to +300°F.
Low temperature polysulfide sealing compound	MIL-S-83318	Intended for quick field repair of other sealants. Fast cure rate at low temperatures.

Table 5-3-1 Common potting and sealing compounds

Corrosion Removal Materials and Equipment

Corrosion removal. When corrosion is detected, corrective action is required. When the corrosion is within repairable limits, specified in the applicable original equipment manufacturer's (OEM) manual or other directive, corrective action should be initiated. This should consist of paint removal as required, cleaning, corrosion removal, treatment and the application of protective coatings and preservation. The mildest method of corrosion removal should always be used. The following paragraphs list some of the methods used on avionics equipment.

Chemical paint removers. Epoxy paint removers may be used to chemically remove paint from metal surfaces. This paint remover should be applied by brush. Care must be observed to ensure that the paint remover does not contact any part of the body. Observe all cautions and warnings. Whenever a chemical paint remover has been used, ensure that the surface is washed with a detergent and water mixture and thoroughly rinsed prior to the application of other coatings. Properly dispose of hazardous wastes generated during the stripping process.

Corrosion removal equipment and methods. The nature of some surfaces, such as chrome, nickel, gold and silver-plated contacts, restricts the use of highly abrasive methods. Tarnish and light corrosion can be removed from such surfaces by hand rubbing with a clean pencil eraser, brushes and nonabrasive pads. Surfaces such as covers, connectors, receptacles, antenna mounts, equipment racks, chassis, etc., may have light to moderate corrosion removed by an abrasive mat or cloth.

Portable mini-abrasive unit. The portable mini-abrasive unit is a hand-held miniature abrasive tool used to remove light corrosion products from small avionics components, such as PCBs, edge connector pins, small avionics structural components, etc. Care should be taken not to remove the thin plating from these surfaces.

Hand-held abrasive tools. The hand-held abrasive tool is used to remove corrosion products from larger components such as avionics equipment structures and housings. Glass beads are generally the specified abrasive material used in hand-held abrasive tools. Contamination of other equipment or components by glass beads can occur if they are allowed to blow freely into the surrounding shop area.

Protective Coatings

Protective coatings (generally paint) are susceptible to damage by handling, accidental scratching and corrosion. The function of boxes, chassis, housings and frames are to enclose, protect and secure the vital internal components of the avionics unit. Therefore, with proper maintenance of the protective coating, the structural integrity and protection of the avionics unit can be maintained.

Painted surfaces. Painted surfaces on avionics equipment will withstand a normal amount of abrasion from handling and hand tools. When the painted surface becomes chipped, scratched, or scuffed the loss of the protective coating allows the base metal to become more susceptible to corrosion. Maintenance personnel should pay particular attention to the use of tools around and handling of, avionics equipment. When protective coatings are properly applied, they will prolong the useful life of the base material by protecting it from corrosion and harmful agents.

Minor paint damage. Minor paint film damage generally occurs when maintenance personnel use hand and power tools on and around avionics equipment. Damage can also occur from equipment handling. The result of this damage is a protective finish that is chipped, scratched, or abraded. Short-term protection includes the application of a water-displacing corrosion-preventive compound. Long-term repair of a damaged protective coating is accomplished by touching up the paint system.

Extensive paint damage. Extensive paint damage requires stripping of the old paint, evaluation for corrosion, cleaning, conversion coating for aluminum and magnesium, priming and the application of one or two finish coats.

When the surface to be painted is contaminated, the paint system will not properly adhere. Almost all paint system adhesion failures such as peeling, flaking, etc., are the result of an improperly prepared (contaminated) surface. Contaminants include oil, grease, dirt, moisture, or defective paint systems (loose or cracked paint). Proper cleaning and surface preparation includes removal of corrosion products and other contaminants.

Encapsulates. Encapsulates are materials used to cover a component or assembly in a continuous organic resin. Encapsulates provide electrical insulation resistance to corrosion, moisture and fungus and mechanically support the components. In avionics equipment, encapsulates are classified as follows:

- Potting compounds used to seal electrical connectors, plugs and receptacles.
- Fungus proof coatings, usually varnishes, used to encapsulate certain avionics circuit components in a thin protective film that is impervious to fungal attack.
- Conformal Coatings used to encapsulate PCBs and modules.

Potting compounds. Potting compounds are used for their moisture-proof and reinforcement properties. They are used on electrical connectors to protect against fatigue failures caused by vibration and lateral pressure at the point of wire contact with the pin. Potting compounds also protect electrical connectors from corrosion contamination and arcing by excluding moisture, stray particles and aircraft liquids. Some of the more common sealants and adhesives are listed in Table 5-3-1.

> **CAUTION:** *Potting compounds are toxic to the eyes, skin and respiratory tract. Skin and eye protection is required. Avoid repeated or prolonged contact. Use only in well-ventilated areas. Keep away from open flames or other sources of ignition.*

Some potting compounds have a history of reverting to liquid form after a year or two. This reversion is highly dependent on the environment of the potting compound. Compounds that revert exhibit a sticky, viscous, oozing consistency that flows out of the connector back shell. In some cases the reverted potting compounds flow around and through pins and receptacles, insulating the connection where continuity is required.

When using potting compounds, the following precautions should be followed:

- Thoroughly clean the area to be potted using an approved solvent.
- Follow manufacturer's instructions when mixing the base compound and accelerator. Substitutions, partial mixing, or incorrect proportions of the base compound and accelerator may produce a sealant with inferior properties.
- Do not mix base compounds and accelerator components of different batch numbers because substantial electrical properties may result.
- Potting compounds may contain small quantities of flammable solvents and/or may release by-products on curing. Observe all warnings and cautions and

use personal protective gear identified in the MSDS and by the potting compound manufacturer.

- Potting compounds that have exceeded their listed shelf life should not to be used unless tested and certified by an approved laboratory.

- Avoid contamination of the potting compound. Do not use masking tape and fiberboard molds around the connector during the potting compound cure. If potting molds are not finished with the connector or are not available, a plastic sleeve should be constructed. The plastic sleeve will aid in forming the potting compound around the connector during the cure.

- Allow potting compounds to cure until firm prior to installing connector or components in equipment.

- Frozen pre-mixed potting compounds should be used as soon as possible after their removal from the freezer or a significant (up to 50%) reduction in working life can be experienced.

- Remove reverting or reverted potting as soon as possible.

Fungus-proof coatings. Fungus-proof coatings should not be applied indiscriminately to all electrical components. Treat only those components that have a specific need or are designated in the applicable manufacturer's service directives. Fungus-proof coatings can, in some instances, be detrimental to the function and maintenance of equipment. A repair activity should re-coat an entire area only when a touch-up of the previously coated component would not provide the required protection.

Sealants. Sealants are another type of protective film used in avionics equipment. Sealants are usually provided in the uncured form as a liquid or paste which solidifies (cures) after application. Sealants can be applied by hand using a as brush or spatula, or by air or mechanically-powered filleting/injection guns. Sealants form a flexible seal in gaps and depressions, preventing moisture intrusion at mechanical joints, spot welds and threaded closures. In addition, they prevent entry of corrosive environments into faying surfaces, fastener areas, exposed landing gear switches and other metal-encased avionics equipment. Sealants function principally as waterproof barriers. When an inspection notes damaged sealant, it should be removed, the area inspected for damage, and new sealant applied.

Sealant application procedures. Pre-clean the surface where sealant is to be applied to remove dirt, grime, etc. Perform a final wipe-down of the area with an approved solvent, recommended by the sealant manufacturer, using a clean cheesecloth.

To prevent sealant from contacting adjacent areas, mask the area to be sealed with pressure-sensitive tape.

Apply sealant primer or adhesion promoter when recommended by and in accordance with, the sealant manufacturer's instructions. Allow the surface to air dry undisturbed for 30 minutes to one hour prior to sealant application.

The proper mixing of the base and accelerator sealant components is essential for proper curing and adhesion of the sealant. Mixing should be accomplished in a clean environment per the manufacturer's instructions. The accelerator and base components should be added together after the proper amounts are determined by weight or volume, then mixed until a uniform color is obtained. Two-part sealants are chemically cured and do not depend on solvent evaporation for curing.

Once mixed, two-part sealants have an application life. The application life of a sealant is the length of time that a mixed sealing compound remains usable in standard conditions of 70°F (24°C) and 50% relative humidity.

Conformal coatings. Conformal coatings are generally two-part coatings that are applied in a thin coat to and over PCBs and other components. Conformal coatings protect PCBs and components from moisture, fungus and thermal shock. They also protect the components from fatigue failure by their encapsulating properties. The coatings are flexible, flame resistant and are suitable for application to PCBs by dipping, brushing, spraying, or vacuum deposition.

Component desoldered. Conformal coatings must be removed prior to component removal from the PCB. The following describes the degree of recommended coating removal and various removal methods.

Remove only as much coating as is necessary to facilitate removal of the discrepant component. Remove the coating from all solder connection areas to allow for desoldering. Figure 5-3-1 shows a typical process for the removal of a conformal coating from a PCB.

> **NOTE:** *Do not attempt to remove all the coating down to the laminate surface. A thin residual layer will not interfere with component removal.*

Removal methods. There are three methods for removing conformal coatings. No one

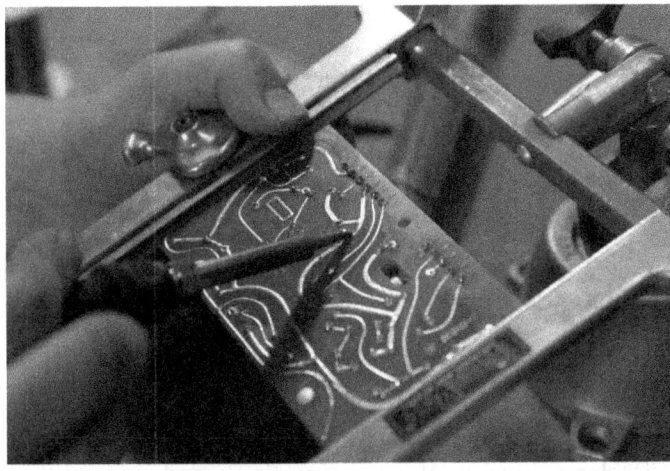

A. REMOVING COATING WITH SOLDERING IRON

B. REMOVING COATING WITH SCRAPER

Figure 5-3-1. Removing protective coating

method or combination is preferred above the others. Methods employed will depend upon coating thickness, circuit density and coating transparency.

Chemical. This method is effective on acrylic resin coatings which may be dissolved by an approved solvent recommended by the coating manufacturer. They may dissolve or swell the silicone resins, allowing removal. They should be applied with a solvent-saturated cotton swab or acid brush.

Controlled Heat. Heat is effective on all resin coatings which will soften or revert (over cure) when exposed to a controlled, localized, low temperature (300°F to 400°F) heat source. The heat source method may also be effective for the release of components from the conformal coating or adhesive bonding of compounds beneath component bodies. This method is recommended on thicker coatings (0.025 inch or greater) because thin coatings offer little thermal protection to the PCB laminate. The use of either low pressure hot air or thermal parting is recommended. The area that is heated should be held to a minimum.

Abrasion. Abrasive removal is effective on all resin coatings. The type of abrasive removal tool and method (motorized or manual) is dependent upon the thickness, hardness, surface contour and transparency of the coating. Motorized abrasive cutting with dental burrs is effective on transparent, thick and hard coatings of polyurethane and epoxy resins. Abrasive discs are effective on all transparent, thin, hard coatings with flat surfaces. Dental brushes are effective on soft to semi-hard coatings with irregular surfaces.

Conformal coating replacement. Preparation of the laminate PCB surface is the most important part of the conformal coating application process. Most adhesion failures are the result of improper cleaning and surface preparation.

1. Prior to the replacement of the conformal coating, all repair action residues (dirt, grime, soldering flux, etc.) should be removed. Clean the PCB laminate surface or component in accordance with the manufacturer's instructions.

2. Dry the PCB or component in a hot air oven.

3. Cleaned and dried PCBs or components should be stored in a clean, static-free plastic bag if the PCB or component is not conformal coated within 30 minutes after reaching the stabilized room temperature. Seal the bag to prevent moisture from forming on the PCB or component.

Selection of a replacement coating. The selection of a replacement coating is based on the compatibility of the replacement coating with the original coating on the PCB or component. Refer to the OEM drawings or repair manual.

Coating thickness. All conformal coatings, with the exception of silicone RTVs, should have the coating thickness applied to 0.003 inch or less, unless directed otherwise by the OEM directive. Applications greater than 0.003 inch will not allow the solvents in the curing coating to evaporate or outgas, thereby trapping the solvents within the curing coating. These trapped solvents can cause bubbles and pinholes in the coating. Silicone RTVs may be applied in a thickness up to 0.008 inch. If multiple coats of any of the coatings are required by direction, ensure sufficient time for the solvent to outgas.

Conformal coating curing. Although the shortest cure time is desirable to speed the production process, the curing time is generally determined by the chemical characteristics of the coating. The curing temperature is limited by the PCB and its components. Conformal coating curing should always be accomplished in a dust-free environment. Elevated temperature curing should be accomplished in a forced-air drying oven at a temperature of not more than 130°F (54°C).

Coating inspection and preservation. The final step in the process of conformal coating replacement is a thorough inspection using an ultraviolet light and a 10x to 15x magnifying lens. Visually inspect the new and original coating for defects such as charring, discoloration, cracking, delaminations, solder flux residue, dry spots, foreign matter, air bubbles and pinholes. If the inspection is satisfactory, the PCB or component should be placed in a polyethylene bag and taped closed.

Figure 5-4-1. Avionics and installed equipment are highly susceptible to corrosion.

Section 4

Treatment of Specific Avionics Equipment

All aircraft electrical/electronic equipment should be opened and inspected for evidence of internal moisture and corrosion on a scheduled basis as determined by the original equipment manufacturer (OEM) or for cause. When corrosion is detected, prompt corrective action is required. Corrective action should include cleaning, corrosion removal, treatment, application of protective finish and preservation where required. Maintenance personnel should always use the mildest method of cleaning and corrosion removal as previously described.

The procedures and techniques for corrosion removal and the restoration of protective coatings are intended to aid the avionics technician in typical repair of specific equipment. In each case, some discretion on the part of the maintenance personnel is warranted. It is important that the maintenance personnel analyze the problem, select the appropriate corrective action and confirm the effectiveness of their corrosion control. The principals covered in Chapter 4, Aging Aircraft, will help to explain the level of vigilance that is required.

Equipment Housings and Mounting Racks

Bilge areas. The bilge area can contain all types of avionics equipment and present a natural sump or collection point for all types of liquids. This is especially true for helicopters. Accumulation of waste, hydraulic fluids, fuel, water, dirt, grime, loose fasteners, drill shavings and other debris is typical. Sump liquids should be pumped or drained from the bilge area whenever discovered. Bilge areas should be cleaned with the proper cleaners and solvents. Maintenance personnel should ensure, prior to installing any avionics equipment in the bilge area, that the area and equipment are cleaned and preserved.

Equipment bays. Avionics equipment bays like the one shown in Figure 5-4-1, and installed equipment are highly susceptible to corrosion. This area is especially corrosion-prone in helicopters and other aircraft and equipment that are cooled with external ram air. Maintenance personnel should perform an inspection of all structures and equipment any time equipment bays are opened.

- Visually inspect the structure, fixed mountings, installed equipment, hard-

Figure 5-4-2. Engine compartment connectors are susceptible to corrosion and heat.

ware and wiring for evidence of corrosion. Particular attention should be paid to dissimilar metal couples.

- Remove corrosion using 320 grit abrasive cloth, or an abrasive nylon mat, in accordance with the OEM's maintenance instructions.
- Clean, rinse and dry the affected area in accordance with the manufacturers instructions and materials.
- After the corrosion is removed, treat all affected aluminum surfaces with a chemical conversion coating.
- Apply primer and topcoat as required
- Apply preservation as required or when environmental conditions dictate.

Engine compartments. Inspect compartment hardware, electrical wire bundles and connectors, terminal boards and junction boxes for evidence of corrosion and damage and repair as needed. Clean and treat corrosion in accordance with the OEM instructions. Figure 5-4-2 is an example of the connectors in an engine compartment.

Frames, mounting racks and shock mounts. Shock mounts and associated hardware on pod or airframe mounted equipment are usually the last items to be inspected for corrosion damage. Figure 5-4-3 shows a moisture entrapment areas on the avionics shelf, equipment racks and shock mounts.

These inspections normally require the removal of shock mounts to facilitate a thorough examination. For frames, mounting racks and shock mounts that are not normally painted, remove corrosion with 320-grit abrasive cloth, or abrasive nylon mat. Pay particular attention to dissimilar metal couples.

> **NOTE:** *The use of dissimilar metals in the selection of screws, washers and nuts should be minimized wherever possible.*

Clean the affected area with isopropyl alcohol. In some shock absorbers, the rubber may swell if saturated with isopropyl alcohol. If the rubber shock mounts do swell, they should return to normal size in a short time after the isopropyl alcohol evaporates.

Frames, mounting racks, shock mounts and associated hardware should be preserved with a thin film of water-displacing corrosion-preventive compound.

Externally mounted equipment. Externally mounted equipment is susceptible to the same corrosive environment as the airframe. Internal and external cleaning techniques are the same as for the airframe except for electromagnetic gaskets, shields, electrical connectors and mating surfaces. Remove corrosion and clean in accordance with the OEM instructions. Conversion coat, prime and paint in accordance with the OEM's maintenance manual and procedures.

Cockpit and control boxes. Cockpits are susceptible to dirt, grime and corrosive attack from cooling air and general human occupancy. Use a vacuum cleaner to clean the

Figure 5-4-3. Potential moisture traps are under and behind avionics racks.

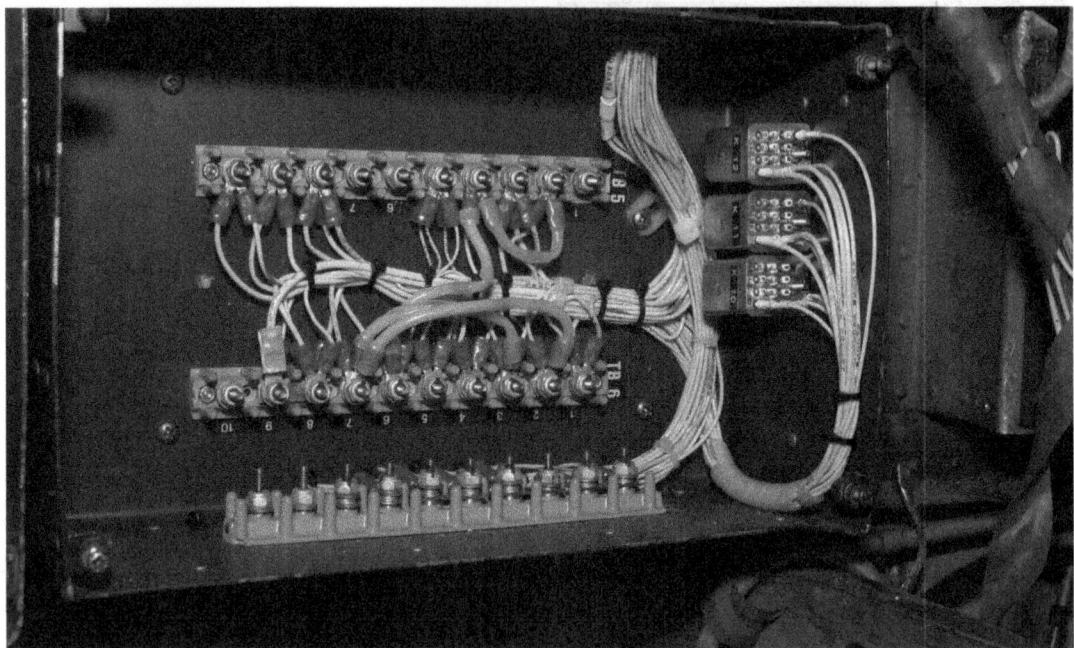
Figure 5-4-4. A junction box with multiple terminal boards and relays.

cockpit area of dirt and dust. Inspect control box units for corrosion and contaminants. Pay particular attention to switches, dials, knobs, electrical connectors, control box and instrument panel mating surfaces and attaching fasteners.

- Remove corrosion products from metal surfaces using an abrasive nylon mat, or an appropriate cleaning and polishing pad.
- Clean the affected area with an approved solvent. Following solvent cleaning, perform a final wipe using a clean cloth and isopropyl alcohol.
- Repaint or touch-up existing paint as required in accordance with the OEM's maintenance instructions manual and procedures.
- Apply a thin film of water-displacing corrosion-preventive compound on exterior metal surfaces and wipe off any excess with a clean cloth. Do not get preservative on acrylic plastic faceplates of instruments.

Terminal boards, junction boxes, relay boxes and circuit breaker panels. Remove covers and access panels, as required. Figure 5-4-4 shows a typical terminal board installation.

CAUTION: *Ensure that all electrical power is disconnected from the aircraft and all electrical systems in the aircraft are deactivated. Disconnect all batteries.*

Inspect, clean and treat external surfaces in accordance with the methods described in the OEM instructions.

- If the application of conversion coating, primer and paint cannot be accomplished, temporarily preserve external surfaces with a thin film of water-displacing corrosion-preventive compound.
- Clean the affected area by wiping or brushing with isopropyl alcohol. While the surface is still wet, perform a final wipe with a clean cloth and allow surface to air dry.
- Preserve internal surfaces of terminal boards, junction boxes, relay boxes and circuit breaker panels by applying a thin film of water-displacing corrosion-preventive compound.
- Avoid applying preservative to relays and circuit breaker contact points.

Metallic equipment covers and housings. Avionics equipment cases, covers, housings and associated hardware may also be exposed to harsh environmental conditions. Inspect, clean, treat and preserve in accordance with the manufacturer's instructions. Hardware (fasteners, clamps, etc.) must be replaced if they are found to be corroded.

Nonmetallic covers and housings. In some cases, avionics equipment, support equipment and general test equipment use nonmetallic high-impact plastic, fiberglass, Kevlar®, or graphite/carbon epoxy covers and housings. These should be inspected, cleaned and the paint touched-up.

Covers and housings should be inspected for structural damage (cracks and delaminations), corrosion around metallic hardware, hinges, latches and damaged paint.

- Repair any structural damage per the OEM's maintenance instructions.
- Clean nonmetallic covers and housings with a solution of fresh water and aircraft cleaning compound applied with a brush or clean cloth.
- Touch-up paint finish as required.
- Lubricate and preserve hinges, latches and unpainted metallic hardware with a conversion coating in accordance with the manufacturer's recommendations.
- Clean shelves, bulkheads and crevices of dirt, lint, debris. Inspect these areas for corrosion and signs of moisture accumulations.
- Pay attention to cracked, chipped, peeling paint and sealants.
- Treat corrosion in accordance with standard procedures for corrosion removal.

Electrical bonding and grounding straps. Bonding and grounding straps used on aircraft and avionics equipment may exhibit galvanic corrosion when not protected. Figure 5-4-5 shows an example of a corroded bonding strap and attach hardware. In most cases the material used for bonding and grounding straps is different than the mating surfaces of the aircraft and avionics equipment.

Antennas, Lights and Connectors

Antenna systems are exposed to severe environmental conditions. Without adequate corrosion protection, these systems can fail because of short circuits, open circuits, loss of dielectric strength, signal attenuation, poor bonding, or electromagnetic interference (EMI). Structural damage to the aircraft may also result. Antennas mounted on the fuselage require openings in the aircraft structural skin to route the various cables to the antenna. The area around the antenna mounting is susceptible to moisture intrusion from rain, runway de-icing fluids and materials, condensation, aircraft wash and internal fluids (i.e., fuel, oil, lavatory and galley products, etc.). Antennas mounted on the lower fuselage are particularly susceptible to corrosion.

Visual inspection. A visual examination of installed antennas and mounting areas can reveal evidence of a corrosion attack. Cracks, splits and peeling of the exterior paint finish and sealant are a good indicator of possible corrosion damage. Evidence of corrosion deposits at the antenna mounting areas is the most obvious indication that an attack has taken place. Examine for a grayish-white to white powdery deposit (aluminum oxide).

Antenna mounting area preparation procedures. When corrosion is visually evident or suspected, corrective action including further inspection by disassembly, is necessary to determine the extent of damage and to prevent further deterioration. Corrosion treatment includes: stripping exterior finishes in the affected area, removing corrosion products, and cleaning and applying surface treatment for avionics bonding. The recommended procedures that may be used on the antenna base and mating aircraft structure for corrosion removal, cleaning and mounting preparation include removing dirt, grime, oil and grease from antenna mating surfaces. Use a cleaning cloth or cheesecloth dampened with an approved solvent. A clean surface will allow proper evaluation of the extent of corrosion damage. Figure 5-4-6 shows a corroded antenna mounting area on a fuselage skin.

- Remove existing fillet sealant from the mating surface of the antenna and aircraft structure/skin with a nonmetallic scraper. Typical dimensions are shown in Figure 5-4-7.
- Remove paint as required from the area surrounding the antenna mounting using epoxy paint remover.
- Thoroughly clean the stripped area with a water-moistened cleaning cloth or cheesecloth.

Figure 5-4-5. Corrosion on a bonding strap can create problems with the avionics.

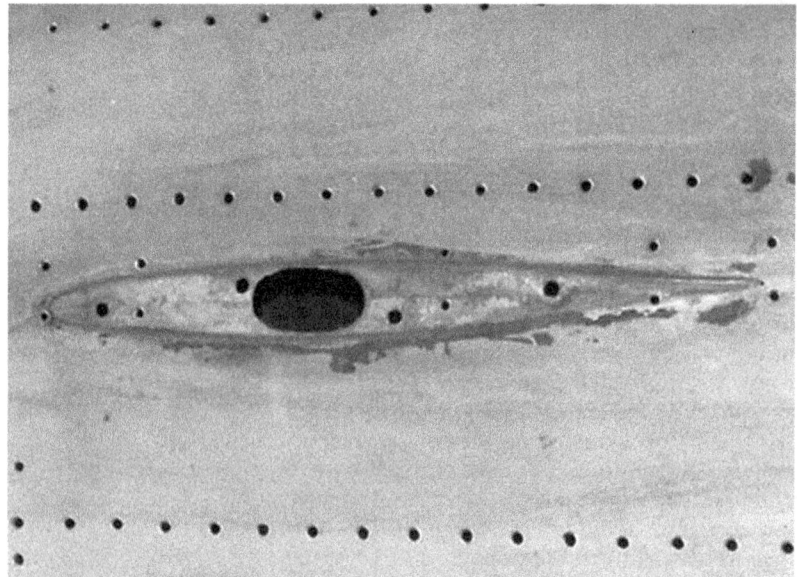

Figure 5-4-6. Corrosion under the antenna interferes with the antenna capabilities.

Installing Antennas

Rigid antenna mounting (without gasket). The mounting bases of rigid antennas vary in shape and size. The following installation procedures are typical and may be used for mast-type antennas (blade, spike, whip base, or long wire mast base) not requiring a gasket. Figures 5-4-8 and 5-4-9 show typical antennas.

Recommended procedures for the application of corrosion prevention measures and attachment of the antenna to the aircraft structure or skin are to remove corrosion from screw fastener countersinks and fastener bore areas on the antenna base in order to provide good electrical conductivity from the base to the screw fasteners.

- Remove corrosion deposits to the limits of the OEM's maintenance instructions with 320-grit abrasive cloth or an abrasive nylon mat.
- Wipe off corrosion removal products using a cleaning cloth dampened with isopropyl alcohol and allow the surface to air dry.
- If bare metal was exposed on the antenna base or aircraft structure/skin, treat the cleaned surface with a chemical conversion coating.
- Install antenna per the OEM's maintenance instructions or apply an even coating of corrosion-preventive compound, to the antenna base and mating aircraft structure or skin. Avoid applying the material in the antenna base fastener countersink areas.
- Position the antenna and install, set and torque the attach fasteners per the OEM's maintenance instructions.
- As an alternative, mix and apply an even coating of polysulfide sealing, to the antenna base and mating aircraft structure/skin.
- Within the working time of the sealant, position the antenna and install, set and torque the attach fasteners per the OEM's maintenance instructions.

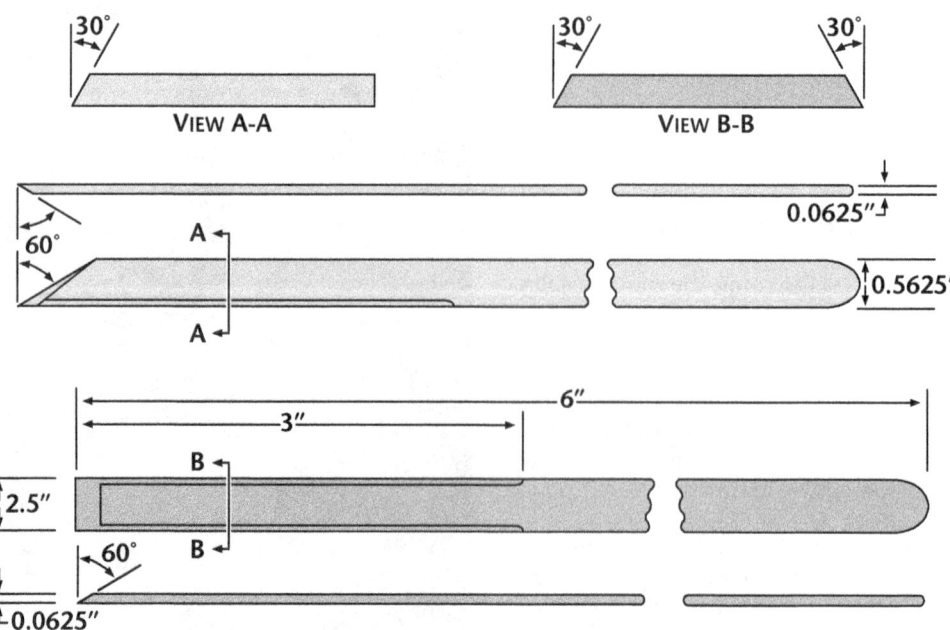

Figure 5-4-7. Nonmetallic sealant removal tools

- Ensure sealant squeeze-out from all sides of the antenna base.
- Conduct an electrical resistance test in accordance with the OEM recommendations.

NOTE: *Mask as required an area just outside of the antenna base on the aircraft skin using pressure-sensitive tape. Remove masking tape within the working time of the sealant. Allow the sealant to fully cure before any flight operations.*

Rigid antenna mounting (with gasket). These procedures are applicable to blade or spike antennas that require a conductive gasket between the antenna base and aircraft skin or structure. The following procedures are recommended for the application of sealant, corrosion prevention and mounting the antenna base:

1. With the gasket in place on the antenna base and the coaxial connector mated, apply a coating of corrosion-preventive compound with a brush.
2. Apply the compound to the skin around the edge of the coaxial cable hole and to the coaxial connector. Ensure the corrosion-preventive compound will not interfere with (insulate) the conductive gasket and the antenna when mated to the fuselage skin.
3. Clean any corrosion-preventive compound from fastener countersink areas with isopropyl alcohol.
4. Position the antenna base.
5. Ensure countersink area is clean under the fastener heads.
6. Set and torque the fasteners.
7. Conduct an electrical resistance test.

Flush or dome antenna mounting. These installation procedures are applicable to flush or dome covered (radome) antennas. These antennas are usually installed on aircraft as part of the primary structure. The radiating elements of the antenna and fiberglass cover are normally individual units. Figure 5-4-10 shows a flush mounted antenna.

The recommended procedures for applying corrosion-preventives to these antennas is to install the antenna in accordance with the OEM's installation instructions.

1. Prior to attaching the cover or dome (radome), spray a coat of water-displacing corrosion-preventive compound on the internal areas of the coaxial and antenna connectors.
2. Shake out the excess.
3. Mate the coaxial connector with the antenna.
4. Spray a coat of water-displacing corrosion-preventive compound on the exterior of the antenna, coaxial connectors and all other exposed metallic hardware.
5. Mount the antenna and secure the fasteners.

Figure 5-4-8. A blade antenna on top of a fuselage

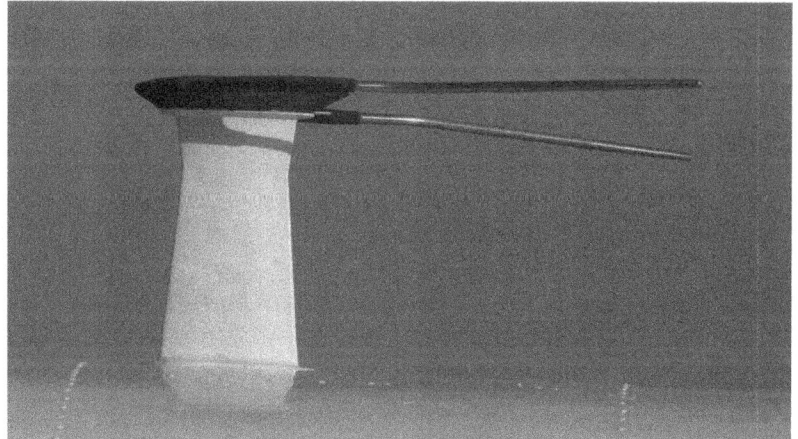

Figure 5-4-9. A V-shaped communication antenna

Figure 5-4-10. A flush-mounted antenna used for navigation

Radar dish antenna corrosion prevention. One of the primary problems related to dish antennas is maintaining the integrity of the protective finish. The protective finish is subject to scratches and chipping from normal maintenance and handling. The metal used in antenna construction is usually aluminum or magnesium. Both metals are anodic to the attached hardware and subject to galvanic corrosion around the hardware. The dish and remainder of the mount are subject to surface corrosion when the finish is damaged. Spot touch-up the paint finish as required in accordance with the OEM instructions. Complete paint stripping, cleaning and refinishing is justified if 20% of the paint finish is damaged. Some radar antennas have a protective finish or covering of Mylar over the base aluminum or magnesium. For these types of antenna dishes, refer to the OEM's maintenance instructions.

UHF/VHF/ADF antenna seating. The location of UHF/VHF/ADF antennas is normally on the lower fuselage. Figure 5-4-11 shows antenna mounting area on a lower fuselage skin.

These mounting locations are particularly susceptible to corrosion because of fluid entrapment in bilge areas. This fluid entrapment is the principal reason for additional maintenance requirements:

1. With the antenna removed, clean the antenna and the mounting location on the aircraft of grease, oil and dirt using an approved solvent.
2. Remove corrosion deposits to the limits of the OEM's maintenance instructions with 320-grit abrasive cloth or an abrasive nylon mat.
3. Wipe clean corrosion removal areas on the antenna and aircraft skin/structure with isopropyl alcohol. Allow surfaces to air dry.
4. Install the UHF/VHF/ADF antennas per the OEM's maintenance instructions.

Long Wire Sense Antenna Corrosion Prevention. The cleaning and preservation procedures for this type of antenna include wiping clean all antenna parts with isopropyl alcohol and allowing the surfaces to air dry.

1. Assemble long wire or DF antenna parts and install in accordance with the OEM's maintenance instructions.
2. After installing and adjusting the tension of the antenna, spray a coating of water-displacing corrosion-preventive compound over the attached hardware. If a new bare wire (no nylon sleeve) was installed, wipe the wire with a clean cloth soaked with water-displacing corrosion-preventive compound.

NOTE: *If the long wire is the stranded type and the corrosion is more extensive than light surface corrosion, it is normal practice to replace the wire. Stranded wire will wick corrosives through capillary action, causing extensive damage that may not be readily apparent.*

Bonding/Ground Connection Electrical Resistance Test

The electrical resistance test is performed after an antenna base is mounted or ground installation is assembled. The test should take place prior to the application of sealant. The test uses an ohmmeter to obtain resistance readings between the antenna base or grounded portion of the antenna and the aircraft skin/structure. It is essential that the test probe be placed against bare metal when taking the readings.

NOTE: *Select a scale on the ohmmeter that will allow a reading of 2.5 mili-ohms. This will ensure maximum instrument accuracy. Proper torque on the connections, good resistance readings and complete seating of the antenna installation are all essential to ensure the antenna will function properly in service. The maximum allowable resistance reading between the antenna base or grounded portion of the antenna and the aircraft skin/structure is 2.5 mili-ohms.*

Avionics Test Equipment

Precision measurement and test equipment is required for testing, troubleshooting and repairing avionics components and systems. This makes the reliability of this equipment in any environment critical to safe aircraft flight and system functions. Aircraft operational requirements often result in short troubleshooting and repair times for malfunctioning avionics systems. Valuable maintenance time is lost if this equipment is not functioning properly. A major source of test equipment malfunction is corrosion on contacts. Corrosion sources that are particularly detrimental to avionics test equipment include moisture and fluid intrusion (rain, condensation, fuel, hydraulic fluid, etc.), corrosive elements in the surrounding atmosphere, malfunctioning or inadequate shop environmental control systems and malfunctioning or inadequate built-in filter systems.

A program of cleaning and corrosion control for the test equipment is part of every avionics shop. The program should be followed to ensure the equipment works properly.

Figure 5-4-11. Belly-mounted antennas corrode faster than other antennas.

Lighting Systems and Assemblies

External formation lights, wing tip lights, rotating beacons and anti-collision lights are highly susceptible to corrosion. The corrosive attack is usually caused by poor seals which allow moisture intrusion from aircraft wash or from the environment in flight. Interior lights and equipment-mounted bulbs are also susceptible to corrosive attack. In most cases, corrosion will be heaviest at the base of the bulb because of the dissimilar metal contact between the bulb and the bulb socket.

> **CAUTION:** *Ensure that electrical power is disconnected from the light assembly prior to corrosion removal and preservation procedures.*

Exterior mounted light assemblies. Remove the light cover assembly and bulb from the socket in accordance with the OEM's maintenance instructions.

- Remove corrosion using an abrasive nylon mat.
- Scrub the affected area to loosen corrosion and contaminants.
- Clean affected area with isopropyl alcohol. Use an acid brush with the bristles trimmed back to assist in cleaning the base of the light socket.
- After cleaning, re-apply isopropyl alcohol to the light socket using a squeeze spray bottle to flush out any remaining residue.
- Wipe the light socket with a clean cloth and allow the area to air dry.
- Assemble the light assembly and touch up exterior paint finish as required.

Interior lights and small equipment light assemblies. Interior lights and small equipment assemblies should be cleaned and preserved in accordance with the OEM and manufacturer's maintenance manuals.

Preserve the exterior of the light assembly using water-displacing corrosion-preventive compound. Assemble light assembly in accordance with the maintenance instructions.

> **CAUTION:** *Allow the solvents in the corrosion-preventive compound to outgas prior to installing the light assembly.*

Relay and circuit breakers. Corrosion (tarnish) removal is required on most types of contacts. Tarnish acts as an electrical insulator. Sliding-type contacts have a self-cleaning action and tarnish removal is not required if a bright surface area is visible. Relay and circuit breaker contact areas are usually plated with a highly conductive metal. Care should be taken to avoid removing this plating. If the plating is removed during cleaning, replace the relay or circuit breaker.

- Heavy corrosion and tarnish may be removed by rubbing with a fine abrasive mat.
- Remove medium corrosion and tarnish by rubbing with a clean pencil eraser.

Figure 5-4-12. A PCB with edge connectors.

- Rinse contacts with cotton swabs moistened with isopropyl alcohol.

- Clean remainder of relay or circuit breaker with an acid brush wet with isopropyl alcohol. Pipe cleaners may be used in hard-to-reach areas to assist in swabbing residue.

- Remove isopropyl alcohol from relays or circuit breakers using a clean cloth and then allow them to air dry.

- Preserve relays and circuit breakers by applying a thin film of water-displacing corrosion-preventive compound to all areas of the relay or circuit breaker, avoiding contact and mating areas.

Edge connectors and mating plugs. Edge connectors on PCBs pose a particular corrosion problem because of the thinly plated surfaces. Most plugs and connectors used in micro-miniature circuit boards are plated with thin layers of gold. This gold is porous and moisture will penetrate to the base metal, causing corrosion. In addition, the very function of cleaning may create scratches in the plated surfaces which will accelerate the problem. Figure 5-4-12 shows the edge connectors on a PCB.

Figure 5-4-13. A connector with plastic inserts in the unused holes.

- Remove corrosion and tarnish by rubbing the affected area with a new pencil eraser. Care should be taken not to remove thinly plated surfaces.

- Clean the contact areas with an approved solvent using an acid brush. Rinse affected area with isopropyl alcohol and allow the component to air dry.

- Edge connectors should be preserved by spraying a thin coating of water-displacing corrosion-preventive compound to both male and female sections of the connector. Wipe off excess preservative with a clean cloth.

Multi-pin Electrical Connectors

Multi-pin electrical connectors require special attention to prevent corrosion and electrical failures, especially when the connectors are in areas that are exposed to harsh environments.

CAUTION: *Ensure that all electrical and hydraulic power is removed from the aircraft or component. Install applicable safety devices. Disconnect all batteries.*

CAUTION: *Cleaning compounds and solvents may react with some encapsulants or plastics used to form wire harness tubing, wire coatings, conformal coating, gaskets, seals, etc. Test on a small area for softening or other adverse reaction prior to general application. A continuity test does not preclude a visual inspection of connectors, because corrosion can still occur outside of pin areas.*

- Protect open connectors with conductive plastic or metal caps.

- Pressure-sensitive tape is an alternate method of sealing connectors if proper caps are not available.

- If connector boots are installed and water intrusion cannot be prevented due to design, a small drain hole (1/4 inch minimum, 3/8 inch maximum) may be incorporated at the lowest point on the connector boot to allow water to drain. This action requires approval of the OEM.

Special attention should be given to connectors using replaceable pins. These connectors use a self-sealing gasket that seals the connector against water intrusion. "Dog bones" (plastic inserts) are used to fill unused connector (pin) cavities. Figure 5-4-13 shows "dog bones" installed in a multi-pin connector. The self-sealing gasket may lose its effectiveness to seal against water intrusion with repeated removal and replacement of connector pins or omission of the "dog bones." The use of potting compounds may be required to prevent water intrusion in extreme cases where the connector cannot be replaced.

Connectors that are susceptible to the same environment as the aircraft wire harness connectors should be treated with the same corrosion removal and preservation techniques. Mounting plates normally contain a gasket that acts as a watertight seal. These gaskets should be inspected each time a connector is dismantled for cleaning or repair.

Corrosion inspection, removal and cleaning. Inspection, removal and cleaning of corrosion on the exterior of electrical connectors should be performed by disassembling the connector backshell, if possible and visually inspect all parts for evidence of corrosion. Figure 5-4-14 and 5-4-15 are examples of disassembled connectors. Extensive corrosion damage may require the replacement of the connector.

1. Remove corrosion by scrubbing with a nonabrasive pad or an abrasive nylon mat, as appropriate. Ensure connector mating surface threads, shell and mounting plate (if used) are cleaned.

2. Wipe off residue with a clean cloth or cheesecloth. Apply isopropyl alcohol with a small brush or toothbrush.

3. Scrub connector mating areas, threads, shell and mounting plates.

4. Remove solvent and residue with a clean cloth.

Inspection, removal and cleaning of corrosion on the interior of electrical connectors should be as follows:

1. Visually inspect all areas of the connector for evidence of corrosion. Extensive corrosion damage may require the replacement of the pin(s) or the connector.

2. Clean internal areas of the connector, wiring and pins with isopropyl alcohol and an acid brush.

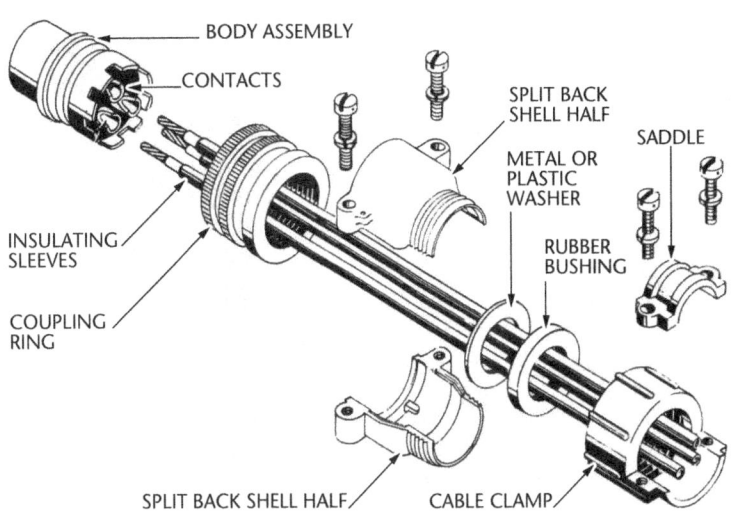

Figure 5-4-14. Installation of Amphenol Class B connector

3. Wipe excess solvent and residue with a clean cloth or cheesecloth. Use a pipe cleaner, as required, to remove solvent from the pin area.

NOTE: *On most connectors it is difficult to clean and remove corrosion from the receptacle (female) contacts. If corrosion is noted, the most practical solution is to replace the pin.*

Sealing connector backshell. Moisture intrusion into a connector often occurs by way of the backshell. This problem is particularly acute after damage to the seal occurs during pin replacement. The backshell may be sealed by verifying that sealing plugs ("dog bones") are installed in unused contact cavities.

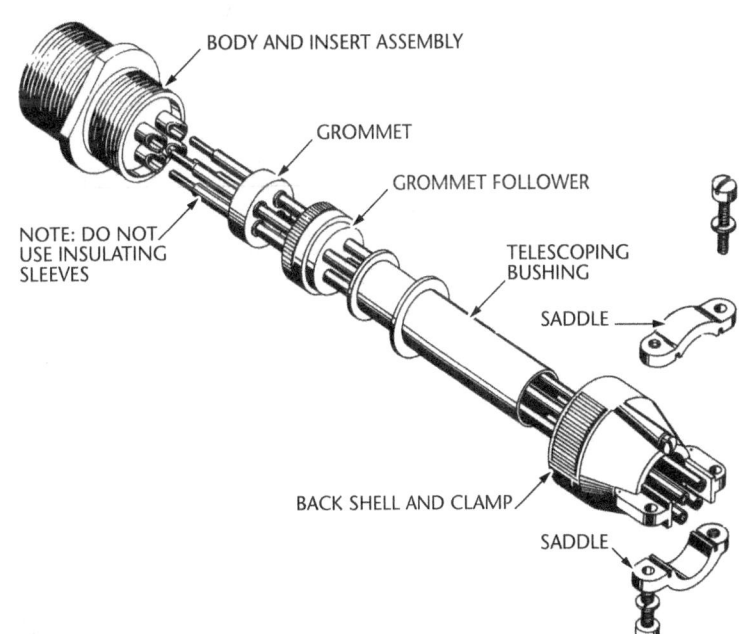

Figure 5-4-15. Installation of Cannon Class E connector

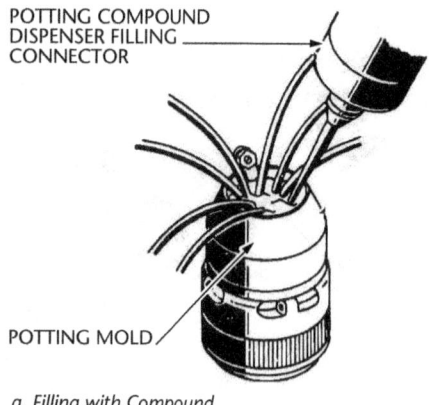

a. Filling with Compound

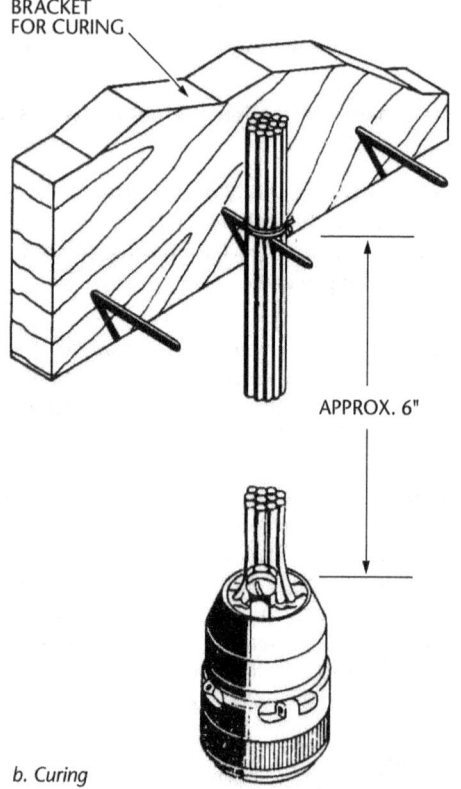

b. Curing

Figure 5-4-16. Filling and curing potting connector

6. Repeat as many times as necessary at different locations around the connector to achieve a sealing compound thickness of 1/16 inch across the entire rear face of the connector.

7. Position connector in the vertical position until the sealing compound sets. The sealing compound will self-level in approximately 15 minutes.

8. After the connector has self-leveled, visually inspect the rear of the connector to ensure a complete seal. If areas are void of sealing compound, add additional sealing compound to entirely seal the back of the connector.

9. If a subsequent repair action requires the replacement of a contact (pin), inject a small amount of sealing compound around the replacement contact to restore the watertight seal.

10. Connectors that are exposed to severe environmental conditions, such as externally mounted connectors and those in wheel wells, bilges, etc., should be taped using an electrical insulating tape.

11. After wrapping the connector and wire bundle with electrical tape, RTV sealing compound should be brushed over the tape.

Water-displacement and treatment. After corrosion removal and cleaning, or any time connectors, plugs, or receptacles are disconnected for maintenance.

1. Apply isopropyl alcohol liberally to internal and external sections of male and female connectors using an acid brush.

2. Mate and unmate connectors several times to clean.

3. Thoroughly rinse the connector with isopropyl alcohol.

4. Shake out excess solvent and wipe connector with a clean cloth.

5. Allow connector to air dry.

6. Spray a thin coating of water-displacing corrosion-preventive compound to the internal sections of connectors, plugs and receptacles. Avoid excessive application or overspray of preservative.

7. If possible, tilt or rotate connector down and around to drain excess preservative.

8. Wipe off any additional preservative with a clean cloth.

9. Prior to connecting the threaded sections of the connector, plug, or receptacle backshell, treat threaded areas with water-displacing corrosion-preventive compound.

1. Remove retainer ring and Mylar tape (if present) from the back of the electrical connector.

2. Slide the backshell and retainer ring in back of the electrical connector up the electrical wire bundle.

3. Tie back shielded wire pigtails, where applicable.

4. Apply RTV sealing compound by inserting the sealant applicator nozzle approximately halfway into the wire bundle at the back of the connector, shown in Figure 5-4-16.

5. Inject RTV sealing compound by squeezing the applicator tube while slowly withdrawing the nozzle from the wire bundle at the back of the connector.

Coaxial connectors. Coaxial connectors require special steps in order to avoid water intrusion. In most cases, water intrusion in fuel/oil quantity indicator and similar capacitance type indicating system connectors will cause erroneous quantity indications on cockpit instruments. Antenna coaxial connectors can generate similar erroneous signals when water intrusion in those connectors occurs. Coaxial connectors should be inspected, cleaned and treated in accordance with standard practices in the maintenance manual.

Static discharge wicks. Corrosion, deterioration, or structural damage to static discharge wicks can result in poor performance from aircraft radios and communication systems, erratic operation of instruments and potential electrical shock to personnel. When damaged or corroded static discharge wicks are found, replace by removing and discarding the old static discharge wicks. Typical static wicks are shown in Figure 5-4-17.

1. Remove corrosion and contaminants from mounting area with a nonabrasive pad.
2. Scrub affected area until all corrosion and contaminants are loosened.
3. Clean and rinse the mounting area with isopropyl alcohol to flush out remaining residue.
4. Wipe dry with a clean cloth and allow the area to air dry.
5. Install replacement static discharge wicks in accordance with the OEM's maintenance instructions.
6. Spray a thin coating of water-displacing corrosion-preventive compound on all metal surfaces and attachment points. Avoid excessive application or overspray of preservative.

Section 5
Electrical Bonding/ Grounding

Electrical bonding provides a low-resistance electrical path between two or more conductive units or components. Grounding is a form of bonding that utilizes the primary structure as a portion (return path) of the electrical circuit. Bonding may serve as one or all of the following functions:

- Provide a common ground for the proper electrical functioning of the units involved
- Provide a path to minimize lightning strike damage
- Prevent the buildup of static potentials that could result in a spark discharge
- Minimize static and stray currents
- Prevent a unit from emitting electromagnetic energy that would interfere with itself or other units
- Shield equipment from outside electromagnetic interference (EMI) sources

The connection of two or more diverse electrical objects often results in a bi-metallic junction that is susceptible to galvanic corrosion. This type of corrosion can rapidly destroy a bonding connection through physical corrosion damage and the loss of the low resistance electrical path if suitable precautions are not observed. Refer to the most current AC 43.13-1 Acceptable Methods, Techniques and Practices Aircraft Inspection and Repair. Aluminum alloy jumpers (bonding straps) are used in many bonding situations. Copper, tin-plated copper and stainless-steel jumpers are most often used to bond together aircraft and component parts made of stainless steel, cadmium-plated steel, aluminum, brass, or other metals.

Where contact between dissimilar metals cannot be avoided, the choice of bonding material and associated attach hardware is important. When selecting materials for the bonding installation, the material(s) that is the most prone to corrosion (anode) should be the easiest and least expensive to replace. At bi-metallic junctions, where finishes are removed to provide a good electrical connection, a preservative or sealant should be applied to the completed connection to prevent corrosion.

Figure 5-4-17. Static wicks are necessary to bleed off static charges that build up on the aircraft.

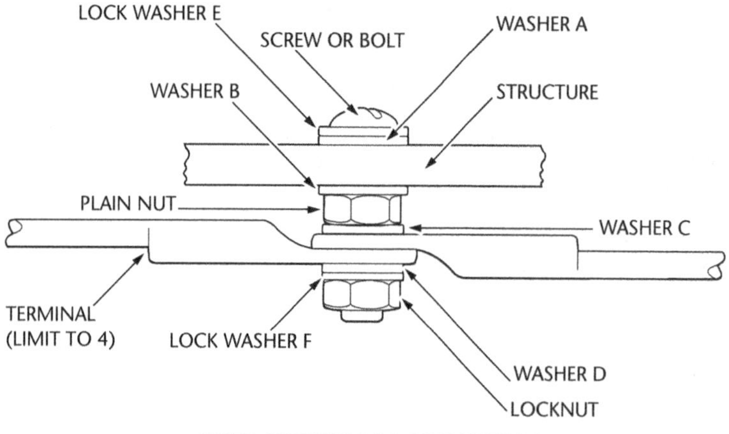

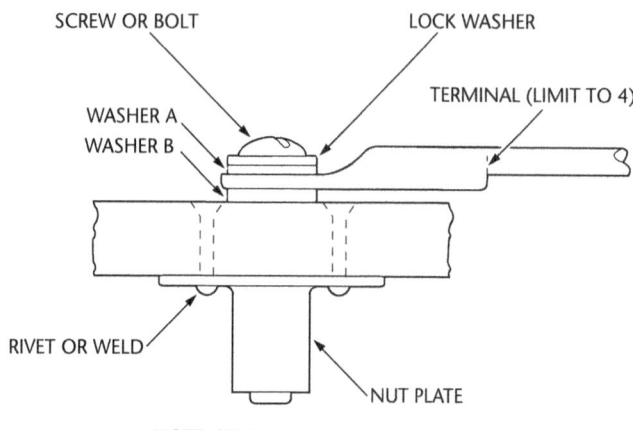

Figure 5-5-1. Stud bonding or grounding to flat surface

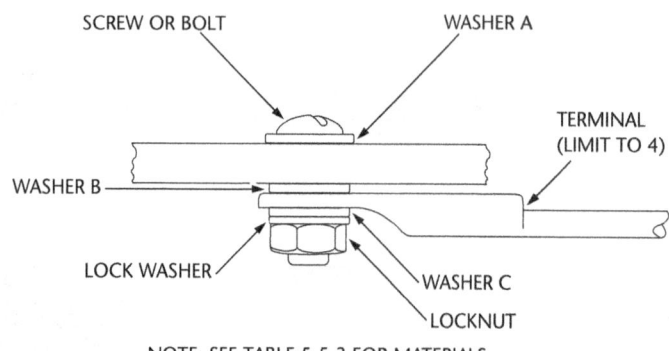

Figure 5-5-2. Plate nut bonding or grounding to flat surface

Figure 5-5-3. Bolt and nut bonding or grounding to flat surface

their mechanical strength, electrical requirements, corrosion resistance and ease of installation. However, when a bonding or grounding connection installation displays evidence of galvanic corrosion after proper assembly, the installation of a sacrificial washer made of an anodic material between the dissimilar materials will allow that anodic washer to corrode.

Replacement of the corroded washer is easy and the least expensive way of repairing the bonding or grounding connection. Figures 5-5-1 through 5-5-3 and their corresponding Tables 5-5-1 through 5-5-3 show proper assembly configurations and list hardware and materials in the order of assembly, depending on the particular metal(s) of the structural and bonding or grounding connection. For example, a proper installation for the "aluminum terminal and jumper" configuration in Figure 5-5-1 shows a bolt secured as a mounting stud for a bonding or grounding connection through a flat structural surface.

The structure in this case is an aluminum alloy and the bonding or grounding jumper, as mentioned, is also aluminum. The attaching hardware are cadmium-plated steel bolts or screws, aluminum washers and cadmium-plated lockwashers and nuts.

Bonding/Grounding Surface Preparation

Surface preparation. Procedures for the preparation of metallic surfaces before mating electrical conductor(s) are to remove all dirt, oil, grease and other contaminants from an area slightly larger than the bonding or grounding connection. The area to be cleaned should be a minimum of 1 1/2 diameters the size of the bonding or grounding connection. Use a clean cloth dampened with an approved solvent.

1. If more vigorous contaminant removal is required, scrub with an acid brush and an approved solvent.

2. Remove, as required, paint, anodic coating or conversion coating film and any corrosion from the attachment area using an abrasive nylon mat. Do not exceed the maximum depth allowed by the OEM maintenance instructions.

3. Wipe area clean with a clean cloth dampened with isopropyl alcohol and allow the area to air dry.

4. Apply chemical conversion coat to aluminum or magnesium as required by the manufacturer.

5. Remove all dirt, oil, grease and other contaminants from the bonding cable terminal with isopropyl alcohol.

Hardware selection. When repairing or replacing existing bonding or grounding connections, follow the original equipment manufacturer's (OEM) parts and maintenance instructions or use the same kind of bonding material and associated attach hardware as the original installation.

The bonding material and associated attach hardware have been selected by the OEM for

STRUCTURE	SCREW OR BOLT; LOCKNUT	PLAIN NUT	WASHER A	WASHER B	WASHER C & D	LOCKWASHER E	LOCKWASHER F
ALUMINUM TERMINAL AND JUMPER							
Aluminum Alloys	Cadmium Plated Steel	Cadmium Plated Steel	Aluminum Alloy	Aluminum Alloy	Cadmium Plated Steel or Aluminum	Cadmium Plated Steel	Cadmium Plated Steel
Magnesium Alloys	Cadmium Plated Steel	Cadmium Plated Steel	Magnesium Alloy	Magnesium Alloy	Cadmium Plated Steel or Aluminum	Cadmium Plated Steel	Cadmium Plated Steel
Steel, Cadmium Plated	Cadmium Plated Steel	Cadmium Plated Steel	None	None	Cadmium Plated Steel or Aluminum	Cadmium Plated Steel	Cadmium Plated Steel
Steel, Corrosion Resistant	Corrosion Resistant Steel	Cadmium Plated Steel	None	None	Cadmium Plated Steel or Aluminum	Corrosion Resistant Steel	Cadmium Plated Steel
TINNED COPPER TERMINAL AND JUMPER							
Aluminum Alloys	Cadmium Plated Steel	Cadmium Plated Steel	Aluminum Alloy	Aluminum Alloy	Cadmium Plated Steel	Cadmium Plated Steel	Cadmium Plated Steel or Aluminum
Magnesium Alloys[1]							
Steel, Cadmium Plated	Cadmium Plated Steel	Cadmium Plated Steel	None	None	Cadmium Plated Steel	Cadmium Plated Steel	Cadmium Plated Steel
Steel, Corrosion Resistant	Corrosion Resistant Steel	Corrosion Resistant Steel	None	None	Cadmium Plated Steel	Corrosion Resistant Steel	Corrosion Resistant Steel

[1] Avoid connecting copper to magnesium.

Table 5-5-1. Hardware for stud bonding or grounding to flat surface (refer to Figure 5-5-1)

STRUCTURE	SCREW OR BOLT; NUT PLATE	RIVET	LOCKWASHER	WASHER A	WASHER B
ALUMINUM TERMINAL AND JUMPER					
Aluminum Alloys	Cadmium Plated Steel	Aluminum Alloy	Cadmium Plated Steel	Cadmium Plated Steel or Aluminum	None
Magnesium Alloys	Cadmium Plated Steel	Aluminum Alloy	Cadmium Plated Steel	Cadmium Plated Steel or Aluminum	None or Magnesium Alloy
Steel, Cadmium Plated	Cadmium Plated Steel	Corrosion Resistant Steel	Cadmium Plated Steel	Cadmium Plated Steel or Aluminum	None
Steel, Corrosion Resistant	Corrosion Resistant Steel or Cadmium Plated Steel	Corrosion Resistant Steel	Cadmium Plated Steel	Cadmium Plated Steel or Aluminum	Cadmium Plated Steel
TINNED COPPER TERMINAL AND JUMPER					
Aluminum Alloys	Cadmium Plated Steel	Aluminum Alloy	Cadmium Plated Steel	Cadmium Plated Steel	Aluminum Alloy[2]
Magnesium Alloys[1]					
Steel, Cadmium Plated	Cadmium Plated Steel	Corrosion Resistant Steel	Cadmium Plated Steel	Cadmium Plated Steel	None
Steel, Corrosion Resistant	Corrosion Resistant Steel	Corrosion Resistant Steel	Cadmium Plated Steel	Cadmium Plated Steel	None

[1] Avoid connecting copper to magnesium.
[2] Use washers having a conductive finish treated to prevent corrosion.

Table 5-5-2. Hardware for plate nut bonding or grounding to flat surface (refer to Figure 5-5-2)

ALUMINUM TERMINAL AND JUMPER					
STRUCTURE	SCREW OR BOLT; NUT PLATE	LOCKNUT	WASHER A	WASHER B	WASHER C
Aluminum Alloys	Cadmium Plated Steel	Cadmium Plated Steel	Cadmium Plated Steel or Aluminum	None	Cadmium Plated Steel or Aluminum
Magnesium Alloys	Cadmium Plated Steel	Cadmium Plated Steel	Magnesium Alloy	Magnesium Alloy	Cadmium Plated Steel or Aluminum
Steel, Cadmium Plated	Cadmium Plated Steel	Cadmium Plated Steel	Cadmium Plated Steel	Cadmium Plated Steel	Cadmium Plated Steel or Aluminum
Steel, Corrosion Resistant	Corrosion Resistant Steel or Cadmium Plated Steel	Cadmium Plated Steel	Corrosion Resistant Steel	Cadmium Plated Steel	Cadmium Plated Steel or Aluminum
TINNED COPPER TERMINAL AND JUMPER					
Aluminum Alloy	Cadmium Plated Steel	Cadmium Plated Steel	Cadmium Plated Steel	Aluminum Alloy[2]	Cadmium Plated Steel
Magnesium Alloy[1]					
Steel, Cadmium Plated	Cadmium Plated Steel	Cadmium Plated Steel	Cadmium Plated Steel	None	Cadmium Plated Steel
Steel, Corrosion Resistant	Corrosion Resistant Steel or Cadmium Plated Steel	Cadmium Plated Steel	Corrosion Resistant Steel	None	Cadmium Plated Steel

[1] Avoid connecting copper to magnesium.
[2] Use washers having a conductive finish treated to prevent corrosion.

Table 5-5-3. Hardware for bolt and nut bonding or grounding to flat surface (refer to Figure 5-5-3)

6. If more vigorous contaminant removal is required, scrub with an acid brush and isopropyl alcohol.

7. Assemble bonding or grounding connection(s) and torque in accordance with the OEM's maintenance instructions, or use Figures 5-5-1 through 5-5-3 and Tables 5-5-1 through 5-5-3 as a guide.

Electronic equipment shock mount bonding and preservation. This type of electrical bonding uses a bonding wire (jumper assembly) or strips of aluminum or copper. Clean base of shock mount and bonding wire (jumper assembly) or strips of aluminum or copper by wiping with a clean cloth dampened with isopropyl alcohol and allow components to air dry.

After assembly of the shock mount and bonding wire (jumper assembly) or strips of aluminum or copper, apply a thin film of water-displacing corrosion-preventive compound over the shock mount and jumper assembly attach area.

Section 6

Electromagnetic Interference Shielding

Electromagnetic interference (EMI) is the presence of undesirable voltages or currents which appear in a circuit as a result of the operation of another electrical source. EMI includes effects from lightning, external radiated radio frequency (RF) fields, or conducted and radiated electromagnetic interference between systems in the aircraft. In this section, the term EMI will include all of these effects. Some examples of EMI-related aircraft system malfunctions are microprocessor bit errors, computer memory loss, audio tones on communication systems and false indications (i.e., alarms, lights, readouts, or power loss).

The results of such malfunctions can severely impact system or subsystem operation. EMI may be radiated or conducted. Typical sources of radiated emissions are radio and radar transmitters, power supplies, generators and transformers. The way in which external EMI intrudes into a circuit is called the coupling mode. Radiated EMI propagates through the air from the source to the victim circuit. An antenna, or a cable which acts as an antenna, couples the EMI to the victim circuit. Conducted EMI is coupled from the source to the victim circuit between common connections, either wiring or metallic structure.

Due to the increase in electronic systems installed in modern aircraft, the importance of many of these systems for flight safety and the decrease in power levels required to upset them, EMI, lightning and high intensity radiated field (HIRF) protection have become an essential part of aircraft design. The use of electrical and electronic systems for full-authority aircraft flight and engine controls has been a significant factor in increasing the importance of aircraft EMI protection.

Aircraft system EMI protection involves the structure around the system, wire routing, shields over the wires, shield terminations, filters inside and outside the equipment, circuit and equipment grounding and circuit design. Most of these protection features rely upon low-resistance and low-impedance electrical bonds for wires, shields and structure, which often include dissimilar metals. For this reason, an understanding of the purpose of these devices, where they may be located and corrosion-control processes are necessary knowledge for the aircraft maintainer.

EMI Protection Requirements

Although electronic and electrical equipment is usually tested individually for lightning, HIRF, radiated and conducted EMI and EMI susceptibility and emissions, aircraft manufacturers should still ensure that the integrated electronic systems on the aircraft will operate successfully in the electromagnetic environment to which the aircraft will be exposed. Figure 5-6-1 and the following paragraphs describe some of the aircraft-specific EMI issues that should be addressed.

Intrasystem EMI requirements. Intrasystem EMI refers to electrical or electronic subsystems within an aircraft interfering with one another. Even subsystems designed to similar emission and susceptibility requirements may have EMI problems when integrated in an aircraft. This may be due to the location of the equipment on the aircraft, frequencies involved, cable routing, or bonding and grounding techniques. All of these factors should be addressed by the aircraft manufacturer in order to ensure a compatible design. In a worst case scenario, when intrasystem EMI cannot be avoided, operation of one system may not be possible while another is operating. Changes to electronic sys-

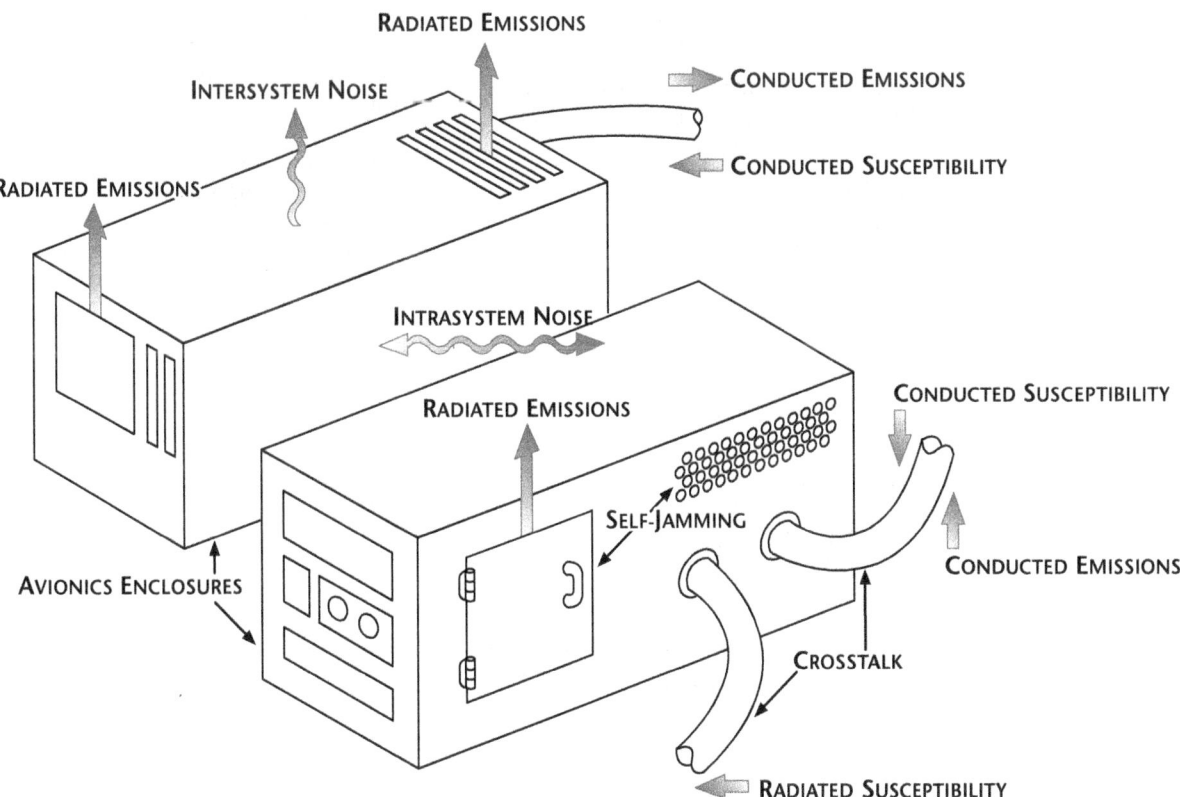

Figure 5-6-1. Common sources of electromagnetic interference (EMI)

tems or installation of new systems can impact Intrasystem electromagnetic compatibility (EMC) as well.

Intersystem EMI requirements. Intersystem EMI refers to electrical or electronic systems external to the aircraft interfering with aircraft systems. These external systems emit various frequencies and power levels which are not always easily predictable. Protection from intersystem EMI can include various types of shielding for aircraft openings and wiring.

Lightning

Effects of lightning strikes. The effects of lightning strikes on aircraft may be classified as either direct or indirect. The direct effects are largely structural. They are the burning, eroding, blasting and structural deformation caused by lightning arc attachment, high pressure shock waves and magnetic forces produced by the associated high currents.

Indirect effects predominantly result from the interaction of lightning fields and currents with electrical equipment or wiring. Fuel ignition may result from either direct arc ignition, or indirect effects creating arcs or sparks. For the purpose of studying lightning effects, aircraft are typically divided into three major lightning attachment zones: direct attachment, swept stroke and conducted currents or transfer. Testing with simulated high voltage strikes on aircraft scale models using various aspect angles can provide confidence in zone assignment.

- **Zone 1** (direct attachment) includes areas where lightning current enters or exits the aircraft.
- **Zone 2** (swept stroke) includes areas directly behind the direct attach points where the established ionized lightning channel is swept back over the aircraft surface as it flies through or away from the channel.
- **Zone 3** (current transfer) includes areas where there is a low probability of direct attachment and which provide a path for current flow through or across the aircraft from entry to exit points.

Design goals. One goal of designing for lightning protection is to provide large conductive areas of structure where high concentrations of current may dissipate rapidly. Small metallic parts, such as control surface actuators, or thin structural members separated from larger metallic areas may be vulnerable to damage. Protrusions such as antennas, probes and light assembly projections are particularly vulnerable to lightning strike attachment.

Discontinuities or nonconductive areas may provide points of entry to the interior of the aircraft. It is important to keep the high current out of sensitive electronics, flammable areas such as fuel tanks, controls and away from personnel. Nonmetallic materials used in aircraft, such as Kevlar, fiberglass, graphite epoxy, Plexiglas, etc., are of particular concern because they do not conduct lightning current-like metallic materials.

Direct effects of lightning may include puncture or burning of these materials. In addition, the high current will seek another, more conductive path. Adjacent wiring or electronics subjected to this high current may be damaged or vaporized.

Some methods of protecting aircraft from the effects of lightning include providing filters or transient suppression for critical circuits; installing diverters to provide a path for current in areas where nonmetallic materials are used; and

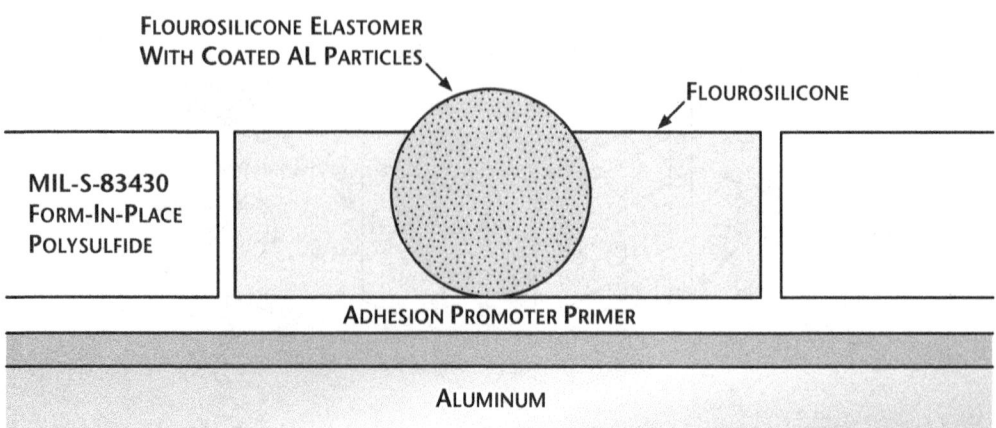

Figure 5-6-2. An elastomeric gasket cross section

bonding of access doors, panels, light fixtures, antennas, probes, landing gear, fuel dumps and fuel vent lines and electronic components (bonding should be low resistance with high current carrying capability for short duration).

Precipitation static. Precipitation static, or P-static, is the triboelectric charging of the aircraft due to flight through dust, ice, rain, sleet, hail, or snow. When precipitation or dust particles contact the metallic or dielectric surfaces, charge separation occurs. This electric charge can cause radio interference through corona discharge from trailing surfaces and streaming from dielectric surfaces or from impact charging itself. Discharges can occur between metallic parts that are not electrically bonded together. In addition to bonding, P-static discharges are often installed on trailing edges and tail surfaces. Static discharges decouple corona discharge currents from antenna fields and permit noiseless discharge to occur. Static discharges should have a low resistance bond to the airframe and a high resistance (between 6 and 200 mega-ohms, depending on placement) from tip to base.

Types of EMI Protection

Structure shielding. Structure shielding is a method of protecting susceptible circuits inside the aircraft from lightning, HIRF and EMI. Metal structures provide a low-impedance path for currents generated by EMI, so that these currents will be minimized on systems and wiring. In addition, enclosed structures, such as the fuselage, provide some shielding for radiated fields.

The principle of shielding is derived from the fact that the total charge completely enclosed by a conductive surface will be zero, regardless of electromagnetic fields external to the surface. The completely closed conductive surface is often called a Faraday cage, named after Michael Faraday, the English physicist and chemist who provided experimental data proving the concept. Of course, no aircraft can be a perfect Faraday cage since there must be openings, doors, windows, vents, etc. The goal of the structure shielding is to seal the cracks or holes in the fuselage to make it as close to a Faraday cage as possible with respect to external EMI that may disrupt internal circuits of the aircraft.

Aircraft structure shielding is more common in new aircraft designs. This is due to several factors, including increased sensitivity and number of electronic components within the aircraft as well as aircraft designs that incorporate composite materials in the structure. These factors lead to greater EMI susceptibility, especially lightning and HIRF (radar, radio transmitters, etc).

Figure 5-6-3. A Hi-TAK® polyurethane conductive gasket installed on an antenna
Photo courtesy of Av-DEC®

Composites are not as conductive as metal and do not provide the same level of shielding, particularly for lightning. The requirement for structure shielding is dependent upon the external electromagnetic environment and the level of protection designed into the wiring and electronic equipment. If structure shielding is required in the aircraft design, it should consider the effects of seams and joints, such as around avionics bays and between the structure and quick access or removable doors, louvers and vents. Gaskets and spring fingers may be avoided during aircraft design because of the maintenance required for these features. The following provides identification of some common types of shielding:

Gaskets. Conductive gaskets may be used to seal access doors and removable panels from EMI intrusion. They should be electrically conductive, fit snugly between the two surfaces of the joint and make good electrical contact between the mating conductive surfaces. A cross-section of a typical gasket is shown in Figure 5-6-2.

Gaskets should be used because of the likelihood of corrosion of the faying surfaces of the antenna and the aircraft skin. Some commonly used EMI gaskets are: elastomers filled with silver or nickel plated aluminum particles; neoprene; silicone rubber bulb seals filled with stainless steel particles and wrapped with stainless steel or Inconel wire mesh; beryllium copper spiral gaskets and polyurethane encapsulated aluminum mesh A polyurethane conductive gasket is shown in Figure 5-6-3.

Gaskets are used most often where frequent access is not required, since repeated compression and decompression of the gasket may result in permanent deformation.

Imbedded metal strips. Aircraft panels made of composite materials may require an occurred metallic plate which makes electrical contact with the embedded conductive composite of the door. Another method uses a bonded foil strip or metallic tape to form a capacitance couple with the conductive composite fibers. Corrosion between the imbedded metal and graphite composites is a particular concern if this technique is used.

Contact strips and spring fingers. Beryllium copper contact strips and spring fingers are used to seal joints between doors and structure in areas where frequent access is required. The strips have evenly-spaced fingers which are mechanically and electrically fastened (metal to metal or capacitance coupled) to the door or structure and pressed against the mating structure or door (metal to metal contact) to provide electrical conductivity across the joint. As with gaskets, contact strips used in joints between composite materials may require a concurred metal plate, conductive foil, or tape for contact with the embedded conductive fibers of the composite material. Again, corrosion between the imbedded metal, spring fingers and graphite composites is a particular concern if this technique is used.

Screens. Screens which cover vents and louvers may be designed to prevent EMI intrusion. The screen mesh should be the correct size to prevent wavelengths of expected EMI from passing through. The screen should also be electrically bonded to the aircraft structure around the entire periphery of the screen.

Bonding. Bonding is the process of establishing a low-impedance (good electrical contact) path between two metal surfaces. The purpose of the bond is to allow radio frequency or lightning current to flow between metallic components, preventing a potential difference or voltage which may result in EMI.

Structural bonding of all parts of the aircraft structure is essential for controlling the conducting path currents associated with lightning, HIRF and EMP. In addition, structural bonding is important to eliminating static charge buildup, which can couple into communication systems. Discontinuities in the aircraft skin (skin joints, access doors, etc.) can create a high-impedance boundary (poor electrical contact) across the joint. Therefore, all discontinuities in the aircraft structure should be designed to provide electrical bonding. Since a low-impedance path is the goal of a bond, the

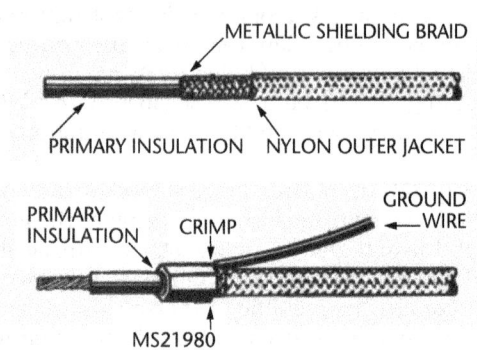

Figure 5-6-3. Wire overbraid provides EMI shielding.

best bond is direct metal to metal contact covering a relatively large surface area and as close as possible to the two surfaces to be bonded. For hinged areas, such as leading or trailing edge flaps, a conductive flexible strap or cable is the best bond that can be provided. The bond areas should be clean and unpainted and the strap should be as short as possible to keep its impedance low.

Bonding of the outer cases of avionics equipment to the aircraft structure is required to ensure maximum operational stability of the equipment and correct functioning of EMI-reducing circuit components, such as filters and shielding. As with the bonding of structural components, the best bond incorporates metal to metal contact covering a relatively large surface area as close to the two surfaces as possible. Beryllium copper pressure-wiper fingers with relatively large surface areas are often used between the equipment and aircraft structure. Flexible conductive straps between avionics equipment and structure may also be used, but the impedance of the bond will be higher.

Aircraft external lighting, antennas, probes, fuel vents and fuel cells are susceptible to lightning and should be bonded to the aircraft structure. Internal circuitry for any of these may require special isolation or high current carrying design requirements.

Electrical circuits, avionics design and aircraft wiring. Most EMI protection is associated specifically with electrical circuits in avionics equipment and the associated aircraft wiring.

The design of aircraft wire routing attempts to avoid both intersystem and intrasystem EMI problems. In some cases, however, physically separating susceptible wiring from other wiring to protect against intrasystem EMI is not possible. In addition, some wiring in wheel wells, engine bays, cockpit areas and other open areas may be exposed to lightning, HIRF, or intersystem EMI.

In these cases the use of metal overbraid over electrical wiring or a wire bundle may be the only method to protect the susceptible wiring. Materials used for metal overbraid vary. Corrosion is a serious problem for overbraid made of ferrous material, particularly in high-moisture areas such as wheel wells. Overbraid of tin-coated copper or Inconel is more commonly used in these areas. Cable overbraid shielding is only effective if the entire cable is completely shielded and both ends have low-impedance terminations.

Splice areas and the area where the overbraid terminates at the connector are the most likely to degrade since these areas require frequent maintenance. Most aircraft wire shields are terminated using pigtails, which may be attached to ground studs or to connector backshells.

Another method, illustrated in Figure 5-6-3, ensures complete shielding using a tightly knitted wire mesh conductive tape which overlaps the overbraid and contacts the connector all the way around the connector circumference. The wire mesh tape is soldered to the overbraid where it overlaps. When wiring must be repaired at the connector, the knitted wire mesh tape can be replaced easily.

The connector itself can be a problem if it incorporates an impedance discontinuity with the wire mesh tape. Connectors used to terminate shields should have good electrical contact between the shield-terminating backshell, the main connector plug and the mating receptacle. Connectors specifically designed to ensure EMI protection may incorporate conductive contacts between the backshell, main shell and receptacle around their entire circumference. An EMI connector made of a composite material is available and has the advantages of being lightweight and noncorrosive, however the conductive plating material is still subject to corrosion.

Grounds. A ground is a common reference for potential in circuit design. In aircraft, the ground for circuits is usually the aircraft structure. Grounding studs are installed in the aircraft near each subsystem to connect ground wires. Similar functions may be grounded on the same grounding stud. Ideally, the ground reference has zero impedance, is at zero potential and conducts zero current. Since no ground plane is ideal, some potential always exists between ground points, with the possibility of undesirable ground currents coupling into and disturbing the circuits.

A good aircraft design minimizes ground currents by keeping ground wires as short as possible, balancing circuits, using twisted signal and return wires to cancel unwanted signals, using coaxial or biaxial lines for RF circuits and employing a number of other techniques. A ground reference also serves to prevent shock hazard and static charge buildup. Good circuit design relies upon dedicated signal and power return wires, so that the aircraft structure is not used for the return circuit, particularly for highly critical or sensitive systems.

Filtering. Inputs and outputs from avionics equipment usually require filters. These filters may provide EMI noise, lightning transient and HIRF suppression. This is best accomplished with filters at the connector internal to the equipment. The purpose of a filter is to exclude unwanted frequencies while allowing transmission of the desired signal frequencies.

Capacitance or resistor/capacitor filters are often adequate for high-impedance circuits, while inductive filters are needed for low-impedance circuits. In some cases, filters may be grouped on a ground plane behind the interface connector and enclosed in a shielded area. Also, in some cases, filter pin connectors can be used, providing significant weight savings and filtering.

Avionics enclosures. Avionics enclosures are designed to minimize conducted and radiated EMI from entering the avionics and to minimize EMI emissions from the avionics. Conductive gaskets may by used in the avionics enclosures, particularly between the enclosure and the connector receptacle installed on the enclosure. In addition, conductive gaskets may be used between access panels on the avionics enclosure, or between ventilation hole screens and the enclosure. The enclosures may also be designed to segregate the filters for input and output wiring from circuit boards and sensitive electronics within the avionics. The enclosures should also provide a means for bonding the enclosure to the avionics rack or structure. This should not be provided by the power or signal returns.

EMI Protection Maintenance

Common failure modes for EMI protection features include breakage, deformation and corrosion.

Breakage. Aluminum foil and mesh used on composite structure panel may be torn or cut. Bonding straps, particularly attached to moveable surfaces, may break from flexing or aerodynamic forces. Shield-terminating pigtails may be broken during connector mating/dismating. Spring finger contact strips, which are typically 0.005 inch thick, seal joints between panels and structure and are easily broken when panels are removed or during equipment removal and installation. Beryllium copper spiral gaskets, which seal joints between panels and structures and beryllium copper

Figure 5-6-4. A visual inspection of the avionics bay is always part of the inspection.

Photo courtesy of Duncan Aviation

pressure-wiper fingers, which bond electronic equipment to aircraft structures, are also easily broken if too much pressure is applied to them. Light weight screens and the stainless steel and Inconel mesh around bulb seal gaskets are easily torn if care is not taken with them. Fortunately, most vent screens are heavy stainless steel and not easily damaged.

Deformation. Connector receptacles attached to structure and brackets may be deformed, or the brackets deformed, if the mounting screws are over-tightened, or if the bracket is not thick enough. Beryllium copper spring fingers can also be bent or deformed such that they do not contact the structure with enough pressure to seal the joint against EMI. This can also be true for bulb seals and conductive elastomer gaskets if they are pressed past the point where they can spring back (compression set).

Corrosion. Corrosion is one of the most common problems associated with EMI protection for two reasons: finishes over metal are often removed during bonding preparation to ensure good electrical conductivity and many conductive materials used for EMI protection system are dissimilar to the aircraft structure. The two types of corrosion which can occur are electrolytic and galvanic. Unfortunately, aluminum, which is used most often in aircraft structure, is higher on the list than materials used in EMI protection devices.

As a result, it is important to inspect areas where EMI protection is installed to ensure structural components are not corroding. In some cases, to protect the aluminum aircraft structure from corrosion, a sacrificial material such as tin/zinc is applied to the structural side of the EMI joint. The EMI gasket or spring fingers contacting this material make the required metal-to-metal contact for EMI protection and at the same time protection is provided to the aluminum structure.

Consequently, corrosion at the EMI joint will be slower and the sacrificial material can be replaced before corrosion attacks the aircraft structure. Because an electrolyte is required for corrosion to occur, environmental seals are often used in conjunction with EMI protection to prevent moisture from contacting the metal-to metal EMl joint. Conductive coatings on aluminum, such Alodyne coatings, should be used instead of nonconductive anodized coating.

Inspection Procedures

Inspection procedures for EMI protection devices and associated corrosion include visual inspections and EMI testing.

Visual Inspections. The condition of avionics enclosures, bonding straps, shields, shield terminations, structural joints, gaskets, spring fingers and conductive coatings on composites may be assessed during visual inspections as shown in Figure 5-6-4. Wherever EMI protection is installed, it is imperative that periodic visual corrosion inspections be performed. Where sacrificial coatings are used, inspection and re-application will be necessary on a periodic basis as well. It may be necessary to remove some EMI protection to inspect aircraft structure for corrosion if there is a history of it or if the aircraft has been exposed to salt spray. Electronic equipment bonding pressure-wiper fingers and bonding straps should be visually inspected when the equipment is removed and replaced. Broken or damaged bonding devices should be replaced. Metal overbraid on wiring in external areas such as wheel wells should be periodically inspected for corrosion. The limits for corrosion on EMI protection features is very dependent on the aircraft and system design and should be specified in the aircraft and component maintenance manuals.

EMI Testing. Some EMI protection failures cannot be detected through visual inspection. An example is wire overbraid which is covered with an opaque jacket. Shield corrosion under the jacket cannot be detected visually. Compression set may occur in conductive elastomer gaskets, on avionics enclosures, where the gasket has been deformed to the point that it no longer seals against EMI. Structural bonding for P-static and lightning protection cannot be visually inspected without major aircraft disassembly. High-impedance bonds between electronic equipment and aircraft structure also cannot be visually detected. Circuit components such as filters should be tested to ensure correct operation. Electromagnetic vulnerability testing, in which the aircraft is radiated with EMI and aircraft electronic systems are monitored for failures, is costly and unlikely to be conducted unless major problems are suspected. Even simpler EMI testing requires some specialized equipment. Consequently, unless in-flight EMI problems are reported, testing will probably not be conducted.

Index

A

acids 1-4
active-passive cells 1-11
aging aircraft
　Figure 11-61 4-2
aging aircraft, background 4-1
aging aircraft, FAA regulation 4-4
aircraft surface cleaning compound 3-7
airworthiness directives (AD) 4-14
aliphatic naphtha 3-6
alkalis 1-4
aluminum and aluminum alloys 3-16
antennas, inspection and repair 5-19
antennas, installing 5-20
aqueous ultrasonic cleaner 5-6
atmosphere, the 1-4
avionics cleaning equipment 5-5
avionics cleaning procedures 5-6
avionics corrosion cleaning facility 5-4
avionics test equipment 5-22

B

bacteria 1-5
best practices 4-13
blasting 3-18
bonding/ground connection electrical resistance test 5-22

C

cadmium- and zinc-plate 3-23
chemical corrosion removal 3-17
cleaning compounds 3-4
cleaning materials 3-6
　aliphatic naphtha 3-6
　methyl ethyl ketone (MEK) 3-7
　safety solvent 3-6
　sodium bicarbonate 3-7
　sodium phosphate 3-7
cleaning procedures 3-7
common corrosive agents 1-4
　acids 1-4
　alkalis 1-4
　salts 1-4
　the atmosphere 1-4
　water 1-4
concentration cell corrosion 1-8
　active-passive cells 1-11
　filiform 1-9
　metal ion concentration cells 1-9
　oxygen concentration cells 1-9
copper and copper alloys 3-23
corrosion
　treatment of
　　blending pits 3-14, 3-15
corrosion, forms of 1-6
　exfoliation 1-8
　galvanic 1-8
　intergranular 1-7
　oxidation 1-6
　pitting corrosion 1-7
　uniform surface corrosion 1-7
corrosion, influencing factors 1-3
corrosion control program 3-2
corrosion detection 1-2
　visual inspection 1-2
corrosion fatigue 1-12
corrosion removal procedures 3-13
　aluminum and aluminum alloys 3-16
　countersunk fasteners 3-16
corrosion removal techniques 3-11
corrosion theory 1-2
　development of corrosion 1-3
countersunk fasteners 3-16

D

damage tolerance 4-2
damage tolerant structures 4-9
detection of corrosion 1-2
drying equipment and procedures 5-8
dust and lint 5-4

E

electrical bonding/grounding, corrosion control for 5-27
electromagnetic interference shielding 5-30
EMI protection
　types of 5-33
EMI testing 5-37
encapsulates 5-13
exfoliation 1-8

F

fastener sealing 3-10
faying surfaces 3-10
ferrous metals 3-21
filiform corrosion 1-9
fillet or seam seals 3-10
flammable and combustible liquids 3-5
fuel tanks 3-10
fungi 1-5
fungus proof coatings 5-14

G

galvanic corrosion 1-8
general aviation aircraft, inspection programs for 4-11

H

hazardous materials 3-5
heat treatment 1-13
hydrogen embrittlement 1-13

I

injection sealing 3-10
insect and animal attack 5-3
intergranular corrosion 1-7
internal stress 1-12

L

lighting systems and assemblies 5-23
lights, inspection and repair 5-19
lubrication 5-9

M

magnesium castings 3-19
mechanical factors 1-11
　corrosion fatigue 1-12
　fretting corrosion 1-12
　heat treatment 1-13
　hydrogen embrittlement 1-13
　internal stress 1-12
　stress corrosion 1-11
metal ion concentration cells 1-9
metallic mercury corrosion on aluminum alloys 1-6
metals, corrosion effects on 5-1
　aluminum alloys 5-2
　cadmium 5-2
　copper 5-2
　corrosion resistant steel 5-2
　gold 5-2

iron and steel 5-2
magnesium 5-2
silver 5-2
tin 5-3
methyl ethyl ketone (MEK) 3-7
micro-organisms 1-5
bacteria 1-5
fungi 1-5

N

National Transportation Safety Board (NTSB) 4-15

O

organic materials, effects of 5-3
dust and lint 5-4
insect and animal attack 5-3
oxidation 1-6
oxygen concentration cells 1-9

P

paint finishes and touch-up procedures 3-10
pitting 3-14
removal 3-14
pitting corrosion 1-7
potting compounds 5-13, 5-14
preservation 3-8, 5-9
preventive maintenance 3-1, 3-2, 5-1
protective coatings 5-13
purple plague 5-3

S

safety precautions 3-13
safety solvent 3-6
sealants 3-8
sea planes, converted from land planes 3-24
service difficulty reports (SDR) 4-15
sodium bicarbonate 3-7
sodium phosphate 3-7
solder flux residue 5-7
solvents 3-5
solvent ultrasonic cleaner 5-6
special airworthiness information bulletins (SAIB) 4-15
stainless steel, treatment of 3-22
magnetic 3-22
non-magnetic 3-22
stress-corrosion cracking 1-12
structural integrity 4-2, 4-5
supplemental structural inspection document 4-8
supplemental structural inspection programs 4-6
supplemental type certificates (STC) 4-15
surface treatment 3-8

T

titanium and titanium alloys 3-23
type certificate data sheets (TCDS) 4-14

U

uniform surface corrosion 1-7

W

water 1-4
water displacement and treatment 5-26
water immersion 3-25
widespread fatigue damage 4-7